Alastair McIntosh is a Scottish writer and campaigner for social justice and environmental sustainability. He holds fellowships at the Centre for Human Ecology, the E. F. Schumacher Society and the Academy of Irish Cultural Heritages at the University of Ulster. In 2005 the University of Strathclyde gave him an honorary post as Scotland's first professor of human ecology. He guest lectures around the world at institutions including the Russian Academy of Sciences, the World Council of Churches, WWF International in their 'One Planet Leaders' programme and, for the past decade, speaking about nonviolence on the Advanced Command and Staff Course at Britain's foremost military staff college.

Praise for *Hell and High Water*

'Provides a chilling clarity . . . McIntosh's excellent exposé might just clear a path out of the darkness' – Paperback of the Week, *The Herald*

'A concise and concerning summary of the current thinking on the science of climate change . . . eloquent and bewitching' – Institute of Environmental Management, *The Environmentalist*

'Thoughtful, incisive and emotionally powerful' – Duncan McLaren, *Friends of the Earth*

'He is at the forefront of an increasingly important exploration of hope in an apparently hopeless ecological situation' – Michael Fordham, *Huck Magazine*

'A valuable insight . . . a fantastically unlikely combination of insights' – John-Paul Flintoff, *TimesOnline.com*

'What's really significant about this book, politically, is that McIntosh has made green living sound attractive . . . He takes a step back from the problem and looks at the causes behind the causes . . . Of genuine international importance' – Roger Cox, *Scotsman*

'He explores the deep order conditions of hope for our planet in the midst of the crisis of global warming. There is no room for a shallow optimism in our present predicament. Hope is a virtue of a different order of magnitude . . . A deep cultural pathology demands a deep cultural psychotherapy' – Professor Emeritus Edmund O'Sullivan, *Resurgence*

'Among climate change books, Hell and High Water is in a class of its own. I know of no one else who has to date presented such a holistic perspective on our collective challenge' – Peter Vido, *www.ScytheConnection.com*

'It's odd that a book of such bright hope should be based on such practical despondency. But then, this lies at the core of his message. He is saying that only when you have stared into that dark place can you find a hope that is real . . . McIntosh offers a soul-based solution' – Vikky Allan, *Sunday Herald*

HELL AND HIGH WATER

Climate Change, Hope and the Human Condition

Alastair McIntosh

BIRLINN

First published in 2008 by
Birlinn Limited
West Newington House
10 Newington Road
Edinburgh
EH9 1QS

www.birlinn.co.uk

Reprinted 2009, 2011

ISBN: 978 1 84158 622 9

British Library Cataloguing-in-Publication Data
A catalogue record for this book is available from the British Library

Typeset by Carolyn Griffiths, Cambridge
Printed and bound by CPI Cox & Wyman, Reading

CONTENTS

ACKNOWLEDGEMENTS

Many people have contributed to the thinking in this book. In particular, I would like to thank my colleagues at the Centre for Human Ecology with whom I have worked for nearly two decades and who valiantly held together an organisational structure where we could mutually explore the science, psychology and spirituality of global problems. I would like particularly to thank my old boss, Ulrich Loening, who taught me so much about the state of the world, and Osbert Lancaster, who has directed the CHE through stormy times. Thanks also to Professor David Miller and his colleagues at the Department of Geography and Sociology, University of Strathclyde, who have provided the study of human ecology with an anchor point amidst their considerable expertise on the politics of globalisation.

For commenting on portions of the manuscript and general brainstorming, I am thankful to my long-standing ecological mentor, Tess Darwin; Michael Northcott of Edinburgh University's New College; Tom Crompton of WWF UK; Henning Drager of Friends of the Earth UK; Zoë Palmer of the CHE; Mike Price of the Doghouse Duo; and both my mother-in-law, Joëlle Nicolas, who pressed me on the importance of *hope*, and my mother, Jean McIntosh, for several of her stories that I have woven in. Detailed technical attention to aspects of the text was most generously given by Myshele Goldberg of the Department of Geography and Sociology at Strathclyde University, Stephan Harding of Schumacher College and author of *Animate Earth*, Iain MacKinnon of the Isle of Skye and the Academy of Irish Cultural Heritages at the University of Ulster, David Cromwell of the Laboratory for Satellite Oceanography at the University of Southampton, Jean-Paul Jeanrenaud of WWF International in Geneva, and Duncan McLaren, Chief Executive of Friends of the Earth Scotland. All remaining errors, and especially all issues of judgement, are entirely my own responsibility.

Throughout the writing of this book I have been kept variously supported, informed and challenged by old friends and acquaintances including George Marshall of the Climate Outreach and Information Network, John Seed of Australia's Rainforest Information Centre, Gehan, Issy and all at the GalGael Trust, James Jones the Bishop of Liverpool, Tom Forsyth of Scoraig, Luke Concannon of the chart-topping duo Nizlopi and George Monbiot of *The Guardian*.

This book would never have come into being without the inspiration of Hugh Andrew of Birlinn Ltd, and my literary agent, James Wills. Hugh was able to see a book in me that I didn't know was there, and his commission and flexibility allowed the blossom to find its own shape. I have particularly appreciated the professionalism of his staff, including my editor, Andrew Simmons, and my patient copy editor, Nancy E.M. Bailey.

Finally, it is my wife, Vérène Nicolas, who has called me constantly back to the mantra of spirituality, and who has nurtured some of the most inspirational ideas that are shared here. Yet again, thank you for being you, dear love.

For Ossian Nicolas McIntosh

Till a' the seas gang dry, my dear,
And the rocks melt wi' the sun!
And I will luve thee still, my dear,
While the sands o' life shall run.

Robert Burns, 1794

INTRODUCTION

Several years ago my widowed and slightly disabled mother moved from the retirement croft house on the Isle of Lewis to the nearby town of Stornoway. Now well into her seventies, she had acquired a cottage by the harbour thinking that 'city life' would make it easier to cope with the wild winter weather.

Tuesday, 11 January 2005 was a tempestuous day even in Govan – the shipbuilding area of Glasgow where I presently live with Vérène my wife. But further out west on the Outer Hebrides, a storm of unprecedented proportions had come in from the Atlantic. Late that evening my mother telephoned. She was coping, but her voice sounded wraith-like and terrified.

Wind speeds are measured on the Beaufort Scale. Francis Beaufort was an Irish admiral who had first gone to sea in 1787. His original scale went up to Hurricane Force 12. Each gradation related to sailing conditions, thus a Force 12, with sustained wind velocities of between 73 and 83 miles per hour, were those 'to which she could show no canvas', and which, over dry land, might cause 'considerable and widespread damage to structures'. On that January night in 2005, winds of 120 mph (200 kph) were recorded near Stornoway. There was also a very high tide and so, combined with the storm surge of water piled up by the tempest, Stornoway's lower-lying streets became inundated by the sea.

As I metaphorically held my mother's hand over the phone,

she described how waves were bursting over the defensive wall across the road. Shovel-loads of stones hailed against her bulging windows. She feared what might happen if the glass gave way. Salty rivulets percolated in around the windowsills and trickled down through the carpets. The whole street was impassably awash. Anybody venturing out would be at peril not just from the deluge, but also from roofing slates flying around like guillotine-edged banshees.

'I'm exhausted,' she told me. 'My strength is almost gone. I've been up and down the stairs for the past two hours, mopping up as fast as it comes in. The emergency services sandbagged my front door, but they can hardly cope and say there's nothing else they can do.'

The crisis subsided as the tide receded, but that night's storm cost the islands millions of pounds in damage. I visited straight afterwards, and three boats were wrecked outside my mother's house, cast up on the rocks almost to the road. In our village of Leurbost, close friends from school days were dealing with roofs ripped from off their weaving and blacksmithing sheds. On a causeway joining South Uist to Benbecula, a family of five in two cars – thought to have been escaping from rising floodwaters inside their low-lying home – were swept away to death. It was the worst natural disaster and the most terrible storm within the islands' living memory.

Scotland's top politicians of the time immediately promised to repair not just 'the infrastructure damage, but also to repair the confidence and the morale of the local community.'[1] Three years later little has been done. The causeways remain dangerous, and the community feels passed by. I couldn't help thinking that if this was the response to a small place at a time of high national prosperity, what would it be like if 'extreme weather events', as they're called by the meteorologists, get much more common? The aftermath of Hurricane Katrina in New Orleans during August 2005 hardly inspires confidence. George Bush had recognised that full recovery may take 25 years and a $6 billion programme was set in place to patch up the flood defences.[2] According to National Geographic News, this was completed to pre-Katrina standards within two years, 'but the system is actually riddled with flaws, and a storm even weaker than Katrina could breach the levees.'[3] One wonders what the chances are of Bush's 25 years ever being reached without a repeat debacle. Already there is anecdotal evidence that the rich are thinking twice about rebuilding in such threatened areas. New Orleans was poor to start with, but as awareness of global warming spreads one can envision the emergence of neighbourhoods both there and elsewhere socially stratified by climate apartheid. The poor will only be able to afford property that is at risk. For the well-to-do, a house on the hill is coming to mean more than just status with a view.

Events like the Hebridean storm and Hurricane Katrina have forced billions of people around the world to start asking questions about climate change. The idea that the Earth is kept warm by a 'greenhouse effect' is nothing new. It was first put forward in 1824 by the French physicist, Joseph Fourier. The possibility that burning carbon-based fuels like coal and oil could ramp this up into global warming was first advanced by the Nobel Prize-winning Swedish chemist Svante Arrhenius. He made some surprisingly accurate calculations as far back as 1896.[4] Throughout the twentieth century the volume of both carbon emissions to the atmosphere and scientific data about them grew exponentially, but it has only been over the past couple of decades that concern harboured by specialists has broken through into public consciousness. Few things shift consciousness like the fear of death or material loss. And when disasters start to hit home, even those whose lives were previously cocooned from the natural world begin to ask questions; questions like:

1. Is the climate undergoing dangerous change?
2. If so, is that change caused by human impact?
3. And if so, can we mitigate the causes and, where necessary, adapt to consequences?

The problem with reactions to specific events like both of the 2005 storms we have been discussing is that there can be no direct proof that they were 'caused' by global warming. The world's weather systems and the variables that make it up are immensely complex. There can only be greater or lesser degrees of probability that any given extreme event is driven by climatic change. After all, every generation will, by definition, suffer its 'worst storm ever' at some point in people's lives. One or two bizarre anecdotes possibly puffed up by the world's mass media don't make for a scientific case that the foundations of the known world have come unstuck. To build a robust body of evidence requires many such anecdotes, accurately measured so as to start comprising a body of data that can distinguish long-term climate change from short-term climate variability and, especially in Britain, from mere *weather*! As it happens, recent data does suggest that the incidence and severity of Atlantic storms is on the rise. In a paper, 'Heightened Tropical Cyclone Activity in the North Atlantic: Natural Variability or Climate Trend?', a team of American scientists found that 'about twice as many Atlantic hurricanes form each year on average than a century ago'. They conclude that this is consistent with the theory that warmer sea surface temperatures associated with global climate change is pumping extra energy into weather systems and thereby upping the ante.[5]

Other data also supports the 'folk memory' of many old people that I grew up amongst on the Isle of Lewis. I must ask my reader to excuse me if I often draw on Scottish examples in this book. It's a question of needing to dig from where I stand, but I hope that the examples I choose will be seen to have far wider relevance in principle. The old Hebridean folks often said that the balance of nature was being upset. They maintained that winters were warmer than they had been early in the twentieth century, and the summers wetter. You can no longer take short cuts safely across frozen lochs in today's winters, and oats will rarely ripen properly. Sure enough, weather station records for the west of Scotland confirm this folk perception. The figures show that between 1914 and 2004, average temperatures did indeed rise by half a degree. Rain, snow, mist and whatever else counts as 'precipitation' rose over the same period by 9.5%. The statistics can be presented with even greater drama if one looks at the data set for 1961–2004. Here the west Scottish temperature increased by fully one degree, and precipitation by a whopping 23.3% – a disproportionately high share of which falls in winter.[6]

There is now a very wide scientific consensus that data like this, drawn from many parts of the world and reflecting a broad range of climate variables, suggests that significant change really is happening to the planet. As the United Nations' Intergovernmental Panel on Climate Change (the IPCC) uncompromisingly put it in their November 2007 report: 'Warming of the climate system is unequivocal, as is now evident from observations of increases in global average air and ocean temperatures, widespread melting of snow and ice, and rising global average sea level.'[7]

We will see later that the vast majority of experts not in the pay of oil companies believe that the primary cause of this warming is carbon emission caused by the burning of fossil fuels. But the implications for our Western way of life – for what it would take to reduce and stop it – are, shall we say, thought provoking. George Monbiot's *Heat: How to Stop the Planet Burning* is an analysis described by Sir John Houghton, former head of the Met Office, the British weather forecasting service, as 'the best book I know . . . broad, balanced and practical'. I have known George for many years and have every confidence in the accuracy of his appraisal. He sums it up in these words:

> By 2030, according to a paper published by scientists at the Met Office, the total capacity of the biosphere to absorb carbon will have reduced from the current 4 billion tonnes a year to 2.7 billion. To maintain equilibrium at that point, in other words, the world's population can emit no more than 2.7 billion tonnes of carbon a year in 2030. As we currently produce around 7 billion, this implies a global reduction of 60%. In 2030, the world's people are likely to number around 8.2 billion. By dividing the total carbon sink (2.7 billion tonnes) by the number of people, we find that to achieve stabilization the weight of carbon emissions per person should be no greater than 0.33 tonnes per year.

> In the rich countries, this means an average cut by 2030 of around 90%. The United Kingdom, for example, currently releases 2.6 tonnes per capita, so would need to reduce its emissions by 87%. Germany requires a cut of 88%, France of 83%, the United States, Canada and Australia 94%. By contrast, the Kyoto Protocol to the United Nations Framework Convention on Climate Change – the only international agreement that has been struck so far – commits its signatories to cut their carbon emissions by a total of 5.3% by 2012.[8]

Other assessments fall into a similar ballpark. For example, Al Gore, whose documentary *An Inconvenient Truth* became an unexpected box office hit, considers that the rich world must cut its greenhouse gas emissions by 90% by 2050. In a detailed study for Friends of the Earth and the Cooperative Bank, Manchester University's Tyndall Centre prescribes a 90% reduction by 2050, but with the majority of this, 70%, needing to be achieved by 2030.[9] To stabilise carbon dioxide levels in the atmosphere close to current levels the IPCC's 2007 report states that global emissions would need to fall almost immediately by 50–85%, but this would still result in a temperature rise above pre-industrial levels of 2.0–2.4°C and a 0.4–1.4-metre sea-level rise.[10] But the IPCC's small print contains a worrying qualification. Their forecasts are based on older models that incorporated 'the sea-level rise component from thermal expansion only'. In other words, the melting of glaciers is not included. As we will see, data has only recently become available suggesting that the icecaps are melting at rates not previously anticipated. There is always an 'inevitable outdatedness' with scientific reports that synthesise vast amounts of published data. We will therefore need to wait until the IPCC's next report in several years' time for a more comprehensive prognosis.

To cut carbon emissions and thereby mitigate the lead cause of global warming means changing our lifestyles, increasing the efficiency of existing fossil fuel uses, or moving to non-carbon energy sources such as nuclear or renewables. In *Heat*, George Monbiot examines British energy use sector by sector. He shows that the necessary changes could be made to stack up, but his findings would have radical implications for how our society is structured. For example, with transportation there would need to be a virtual end to air travel and a heavy curtailment of car use. Even high-speed trains are not efficient enough. George considers that only a massive expansion of bus routes could deliver what is needed. I can almost feel my readers groan! But this is precisely what makes some of George's words so deeply important. It begs the question as to whether the way we look at this whole issue needs to shift ground. He writes:

> Most environmentalists – and I include myself in this – are hypocrites . . . I would like to believe that the changes I suggest could be achieved by appealing to people to restrain themselves. But though some environmentalists, undismayed by the failure of the past forty years of campaigning, refuse to see it, self-enforced abstinence alone is a waste of time . . .
>
> I have sought to demonstrate that the necessary reduction in carbon emissions is – if difficult – technically and economically possible. I have not demonstrated that it is politically possible. There is a reason for this. It is not up to me to do so. It is up to you . . . The campaign against climate change is an odd one. Unlike almost all the public protests which have preceded it, it is a campaign not for abundance but for austerity. It is a campaign not for more freedom but for less. Strangest of all, it is a campaign not just against other people, but also against ourselves.[11]

So here is the real challenge of climate change. It whisks us up in a whirlwind and throws us down against . . . ourselves. That is why the central thesis of this book is that climate change cannot be tackled by technical, economic and political measures alone. Those things are all important, but in addition and perhaps most important of all, we have to look at ourselves. We have to address not only the outer world of atmospheric science, economic imperatives, and realms of political possibility, but also the inner world of psychology and, I will suggest, spirituality. The bottom line and top priority is that we must get to grips with the roots of life and what gives it meaning.

In attempting so to do I want to stand, if I may, on the shoulders of people like George Monbiot. I will largely take the findings of *Heat* and similarly carefully researched texts as a given. There is no point in writing yet another book about climate change when I am not a climate scientist. As such, the shorter part of this book, Part 1, will merely give a summary of the science and the politics. In Chapter 1, I explore the difficulty of knowing what to think about a complex scientific debate with conflicting media voices. Chapter 2 looks at global climate change scenarios and includes a short case study of Scotland. Chapter 3 summarises the technical options to mitigate climate change. And most dismally of all, Chapter 4 looks at why our hedonistic democracy is so impotent in making changes that, actually, need to start within each one of us.

Thus far my material is not distinctive and readers who are already well-versed in climate change debates may wish to skip or just skim over Part 1. In Part 2, my contribution attempts something different from the usual take on global warming. My thesis is that the most galling aspect of the problem is driven not by fundamental human *needs*, but by manipulated *wants* that find expression in consumerism. To mitigate climate change and even to adapt to its consequences without losing our humanity, there needs to be a radical reactivation of our inner lives. That is not something that we can achieve entirely on our own. It also requires change across society – perhaps even what I describe as 'cultural psychotherapy'.

Part 2 of the book therefore explores the history of and the prognosis for the human condition as it relates to environmental impact. In Chapter 5, I show that the ancient world of the Sumerians, Hebrews, Greeks and Romans displays an astonishing perspective on how the human condition reflects itself in the condition of the Earth. The ancients equated hubris or excessive pride with violence and so, with the destruction of nature. Their moral analysis fits even better to our present day condition than it did to their own.

Chapter 6 suggests that as modernity took root in the West, culturally embedded violence damaged our capacity to develop and sustain a rich inner life. Rather than evolving a healthy balance between our inner and outer lives, Western societies have been turned inside out. It shows especially in the faces of some politicians and celebrities. Our outer lives are hyperactive and there's a corresponding emptiness, even a deathly nihilism, at the core.

Chapter 7 argues that this deficiency of inner anchoring has rendered us vulnerable to colonisation by marketing that has pushed consumerism by generating wants. As we lost touch with inner sensibility our psyches – our totality in body, mind and spirit – became open to hijacking by carefully honed tools of motivational manipulation. Inner climate affects outer climate because inner hubris drives outer hubris in a spiral of mindless economic frenzy. As Leonard Cohen puts it:

> Things are going to slide, slide in all directions
> Won't be nothing
> Nothing you can measure anymore
> The blizzard, the blizzard of the world
> has crossed the threshold
> and it has overturned
> the order of the soul[12]

Chapter 8 suggests that if we want to tackle the deep drivers of consumerism and so tackle the roots of climate change, we need to call back the soul. This means setting aside delusions of mere optimism about the future and blind faith in technical fixes, yet paradoxically, deepening our capacity for hope. It means finding the courage to face death and open the heart to love. It means being prepared to be surprised by potential depths of being of which we might previously have been unaware.

Lastly, in Chapter 9 I tentatively suggest twelve steps by which we might work to re-ground the human condition in what it can mean to be most deeply, and beautifully, human. This means working towards a psychotherapy of the soul – a deep healing of what has gone wrong or never properly developed – that is not just individual, but cultural.

My readers should know that I am painfully aware, as will

be explored further in the Afterword, that this is an uncomfortable and also an unfinishable book. It may disappoint, for I have no easy or adequate remedies for global warming. While I try to be careful not to play up people's fears (and some would say I play them down too much), I cannot say that I am optimistic about saving some of the things that are most familiar and loveable in this world. And yet, my position borders on the perverse. I perversely hold out hope for humanity, not in spite of global warming, but precisely because it confronts us with a wake-up call to consciousness. Answering that call of the wild to the wild within us all invites outer action matched by inner transformation. This book takes an exploratory walk on that wild side.

Credit Crunch Postscript to the Second Edition

As this book goes to reprint within its first year, the science requires no significant revision. Suffice to cite the Copenhagen Climate Change Congress of March 2009. Here 2,500 scientists concluded: 'Recent observations confirm that, given high rates of observed emissions, the worst-case IPCC scenario trajectories (or even worse) are being realised.'

Meanwhile, in London in April 2009, the G20 sought to re-inflate the same old economy with 'sustainable global growth'. But as the UK Government's Sustainable Development Commission said the same week, 'The myth of growth has failed us.' To raise 9 billion people up to OECD levels by 2050, according to its *Prosperity Without Growth?* report, would require a 15-fold economic expansion. While economists dream on, the ecology unravels.

Everything said about ancient hubris here in Part 2 applies directly to the economic crisis. Both the credit crunch and the climate crunch have the same origin. Our resultant predicament is like a tangled ball of string. Pull on any end, and all connects – far and wide, outer and inner, ever tighter. It's grim, but it's exciting: for what it demands of us collectively today is nothing less than visionary depth – the courage of getting more real.

Part 1 – Climate Change

Chapter One

NULLIUS IN VERBA

Wishing to dig from where I stood when I started the research for this book in the autumn of 2006, I ran an internet search on the keywords 'climate change Scotland'. The first link that jumped out was a sponsored one – in effect, a paid advertisement. Bold lettering shouted: **'Climate Change is serious.'**

What was this? An environmental NGO with an over-resourced capacity for spitting out headlines? Not so. Here was the Scottish website of The Royal Society.[1]

Which royal society? Like the letterhead of a stately mansion house that omits anything so common as the street name and number, here was an institution with such a sense of its own prestige that it feels able to dispense with the full title: The Royal Society for the Improvement of Natural Knowledge.

Created by Royal Charter in 1660 and with offices in both London and Edinburgh, the Royal Society has a Latin motto: *Nullius in Verba*. It means, 'On the words of no one'. In the vernacular: 'No bullshit.' Ideas must stand on their own feet or else fall down. Luminaries like Sir Isaac Newton, Sir Christopher Wren and The Lord Kelvin were foundation stones of this august body, but their recognition lay in the strength of their contributions and not because they were time-servers or the son of so-and-so. 'On the words of no one' . . . And so going, or gone, were the days when an argument about the nature of reality could be clinched merely from personal authority or, for that matter, from religious dogma. In was to become experimental method as the measure of all things.

The way for this had been paved by early modern scientific philosophers and especially by Sir Francis Bacon, England's Lord Chancellor under King James VI and I. In 1626 Bacon had published a utopian science fiction, *New Atlantis*. It picked up from where Plato had laid his pen down after describing the technically advanced mythical continent of Atlantis. Plato's Atlantis, as we will later see, succumbed to hubris and so to environmental catastrophe under the wrath of the gods. Bacon, copying the technique used by Plato in *The Critias* leaves his account of Atlantis unfinished. It ends abruptly with the sentence: 'The rest was not perfected.' But this was not before Bacon had set out a scientific new world order complete with flying machines, submarines, animal experimentation for medical research and what we might today call genetic engineering for agricultural improvement. At the end of the day Bacon himself became the victim of one of his own experiments. He caught a chill while outside stuffing a chicken with snow in order to study the effects of refrigeration! That was the end of Sir Francis. But in championing the idea of experimental method he had greatly advanced *empiricism* – the idea that, to be valid, knowledge must proceed by observation, experiment, and building up a body of tried and tested *evidence*.

Through such science, modernity as we know it emerged. Historians vary on the timing, but for our purposes we can say from about the early seventeenth century onwards. Reason was its guiding light, and so the seventeenth century became known as the 'Age of Reason', melding into 'The Enlightenment' of the century that followed. What took place over this time was a gradual cultural shift from values to facts. Facts alone were to be sacred. All else was subjective opinion. The emerging paradigm could be summed up in the expression: 'If you can't count it, it don't count!' Previous schools of philosophy had built up elaborate structures of belief from first principles.[2] They drew on notions considered 'metaphysical' – beyond the physical realm. These were ideas about underlying 'essence' and 'being'. They were articulated in assertions about 'God', 'the Prime Mover', 'natural law', 'soul' and the entire 'superstitious' – as it was coming to be seen – superstructure that encrusted them. Scotland was particularly noted for such arcane speculation, earning it the sometime designation 'metaphysical Scotland'. But now, all over Europe, the Enlightenment was bringing in a new broom to the musty metaphysical cupboard. So far so good. And the science that came out of it created the affluence without which many of us would not be here and living in such comfort today. What's more, there'd been a bad smell in that old cupboard for a long time. It needed a cleaning out.

But a broom readily becomes a thrashing stick. By the mid nineteenth and early twentieth centuries, the white light of reason had blinded our eyes to the baby that, arguably, had always been in the metaphysical bathwater. The cutting edge of empiricism had sharpened into a particularly deadening expression of materialism, known as 'logical positivism'. Positivism is another word for empiricism – it is the notion that *only* 'positive', or evidence-based conclusions about reality are valid. Logical positivism was to become the conventional wisdom of the dominant group in British and American universities during the twentieth century. It could be colloquially summed up in the words: 'If you can't count it, it don't count; and if you can't kick it, you can't count it.' Today we might consider it to have been a fallacy of misplaced concreteness. The very means chosen to appraise validity renders invisible key parts of what we might want to look at. Subtlety gets kicked to death. Logical positivism thereby collapses the world and the human being into a shrivelled and shrunken parody of its own stifling worldview. Meanwhile the French postmodernists went to the other extreme, deconstructing the presumed validity of all things positive. But that need not concern us at this point.

The most uncompromising voice of the empirical movement was A.J. Ayer, author of *Language, Truth and Logic* (1936) – probably the single most influential work of philosophy to emerge from England in the twentieth century. Ayer defined positivism as being that which makes sense of the world because it is of the senses. It thereby invalidates metaphysics because 'the utterances of the metaphysician who is attempting to expound a vision are literally senseless'.[3] In saying this he constructed the perfect Catch-22 argument; one that completely denies the worth of anything coming from the inner life. Remember Catch-22 from Vietnam days? To be in the army you had to be insane; but the only way out was to plead insanity. Logical positivism said that in order to be known and accepted as valid, things had to make 'sense'; but anything intangible like vision, dreams, poetry, story or myth couldn't possibly make sense because they failed what I will call the kickability test.

Science here was moving a long way from the intentions of the original founders of the Royal Society. Science, or at least some of its prominent proponents, was out not just to bring a clean sweep to the metaphysical cupboard, but to knock it 6 feet under. For metaphysics read 'God' in shorthand – the ultimate ground of being – and the trouble with burying God is that the Holy Ghost, so to speak, has a haunting tendency to keep on coming back! Cue Professor Richard Dawkins, whose chair in the Public Understanding of Science at Oxford University is funded by the billionaire Microsoft whiz-kid Charles Simonyi – a man who, in 2007, became one of the world's first space tourists.[4] For Dawkins it's not enough for science to be neutral about the possibility of transcendent metaphysical realities such as he incorporates into the catch-all expression 'God'. Only the wooden stake through the heart treatment will lay the Ghost sufficiently to rest. As he surmises in his brilliant, witty, but ultimately empty book, *The God Delusion*: 'I am not attacking any particular version of God or gods. I am attacking God, all gods, anything and everything supernatural, wherever and whenever they have been or will be invented.'[5]

That, then, is a highly potted and therefore somewhat cavalier history of modern scientific thought. There are plenty of scientists who would see things in other ways and would give metaphysics their place, but frankly, they're not the ones that produced a book that cleared two-thirds of a million copies in its first year on the shelves. Whether we agree with Dawkins or not, he has touched a cultural nerve. Whether we like it or not, he surfs the leading edge of Ayer's positivistic wake, giving a particular spin and thrust that arguably distorts the original rich diversity of Enlightenment ideas. Such is the present-day 'where it's at' of a worldview that has emerged from the minds of some of the men who built The Royal Society for the Improvement of Natural Knowledge. And that is what's so fascinating when that self-same natural knowledge now points to the possibility of limitations or contradictions within its own worldview. As we will see later in this book, unpacking these will take us into territory that becomes, well, metaphysical. The Royal Society today almost hints at such an evolving direction itself! As I wrote these words, I was thrilled to see that the last of its five strategic priorities for its 350th anniversary in 2010 is to 'Inspire an interest in the joy, wonder and fulfilment of scientific discovery'. A philosopher like Dawkins would not disagree with those words. He too borrows such language. But it is borrowed, albeit without the grace of acknowledgement, from the metaphysical realm. It implies human capacities that transcend mere logic. Indeed, it comes very close to religious language.

Many organisations have said that **'Climate Change is serious'**. In the context of the history of ideas just outlined, what causes me to sit up and take note is that none other than the Royal Society – the keynote of scientific caution – has now raised its voice in the same tone.

* * *

The webpage that opened in late 2006 from Google's Royal Society link turned out to be timely and prophetic. It said: 'It has become fashionable in some parts of the UK media to portray the scientific evidence that has been collected about climate change and the impact of greenhouse gas emissions from human activities as an exaggeration.' It then went on to lay out what it called 'twelve misleading arguments . . . put forward by the opponents of urgent action on climate change

. . . highlight[ing] the scientific evidence that exposes their flaws.[6]

I say that these words about media portrayals proved 'prophetic', because in March 2007, just a few months later, something happened in British society that placed the Society's unease in sharp relief. Indeed, it prompted the Society further to sharpen up its web statement into a pithy six-point riposte that went live in April that year. The trigger was a TV programme broadcast by Channel 4 that got the whole country debating climate change. Most people, on both sides of the argument, felt suitably outraged. It became the main talking point of chat shows and newspaper columns for at least a fortnight.

What had happened was that in a TV programme called *The Great Global Warming Swindle*, producer Martin Durkin persuasively suggested that the public had been conned about climate change. The culprits, he said, were a curious combination of besandaled hippies, grant-grubbing scientists, and Margaret Thatcher!

Durkin's documentary had what looked like an impressive cast of scientists. It opened to buccaneering music and headline captions saying:

THE ICE IS MELTING
THE SEA IS RISING
HURRICANES ARE BLOWING
AND IT'S ALL YOUR FAULT
SCARED?
DON'T BE
IT'S NOT TRUE

Channel 4's official website heaped it on with corresponding vim and vigour. Here was a message that everyone yearned to hear. Amidst ads for Nintendo and the House of Fraser, potential viewers were asked:

> Are you green? How many flights have you taken in the last year? Feeling guilty about all those unnecessary car journeys? Well, maybe there's no need to feel bad.

> According to a group of scientists brought together by documentary-maker Martin Durkin, if the planet is heating up, it isn't your fault and there's nothing you can do about it.
>
> We've almost begun to take it for granted that climate change is a man-made phenomenon. But just as the environmental lobby think they've got our attention, a group of naysayers have emerged to slay the whole premise of global warming.[7]

The documentary itself presented an impressive array of interviews and graphs to suggest that the public have been fed a pack of lies. Global warming, it maintained, has nothing to do with anthropogenic causes – that is, with the notion that greenhouse gases released to the atmosphere by *people* are the problem. Rather, the cause is all down to solar activity. The sun does its stuff in mysterious ways that come and go according to natural cycles and these affect the Earth's climate. There's nothing we can or should do about it except lie back in the deckchair, luxuriate on the beach, and soak up the sunshine.

* * *

Before taking Durkin further, let's look at what it is that makes it possible, in the first place, for human beings to lie back on this planet and enjoy the sunshine. What made it possible at all for advanced life to have evolved here?

The answer is green leaves – along with a few other organisms like algae and phytoplankton. These are the photosynthesisers that biochemically capture the sun's energy, producing sugar. It led the pioneering Scots botanist and human ecologist, Professor Patrick Geddes, to coin the expression 'By leaves we live', because our entire food chain is so driven.

In photosynthesis sunlight converts six molecules of water and six of carbon dioxide into one molecule of sugar and six of oxygen. Peer into a pond on a sunny day and it can be watched happening. Tiny strings of bubbles will be seen rising from the leaves of water plants. It's beautiful to observe. Here we see the very air that we breathe being replenished in its goodness.

Oxygen comprises about a fifth of the atmosphere's composition. If photosynthesis stopped it would eventually be all breathed up, burnt up in fires, or consumed in natural chemical processes such as the rusting of metals and break-down of rocks. Without photosynthesis, animal life, including our own lives, would slowly suffocate. The sugar produced as photosynthetic product is the basic chemical building block by which all life is energised. The starch in the potato and the sugars in fruit are so made. The malt that distils into whisky is sunlight. The lion that mauled David Livingstone was, ultimately, solar powered: it got every ounce of its spring from the sun's photons 'captured' by leaves of grass, and stored in the fat of the game it would have gorged on.

Our bodies, therefore, owe their entire energy and oxygen supply to photosynthesis. But even more than that, the very amenability of planet Earth is now believed to be a consequence of life creating the conditions for its own flourishing. Back in the 1960s, when the English scientist James Lovelock was asked by NASA how they might predict whether planets might have life on them, he suggested analysing whether their atmospheres were, in principle, capable of supporting life as we know it. Lovelock realised that our own planet can only support advanced life because more simple forms have created the preconditions. Marine life, for example, captures carbon dioxide (CO_2) and ties it up in sea shells. This 'pumps down' the greenhouse gas and 'fixes' it, eventually forming rocks like limestone and oil shale. A certain amount of greenhouse effect is necessary to keep our planet sufficiently warm for life to flourish. As such, we need some CO_2 to be in the atmospheric mix. But Lovelock saw that without life constantly pumping down excess carbon to become ocean sediments and eventually rocks, the Earth would have had far too much atmospheric carbon and would therefore be a hot and inhospitable place. Early life forms – mainly microscopic ones – thereby tamed the Earth for us. A wonderful self-regulating process is kept in place that maintains equitable temperatures.

During the late 1970s Lovelock named this process Gaia after the ancient Greek goddess, a divine personification of the Earth. At first many in the scientific mainstream marginalised his views, though the Royal Society had made him a Fellow in 1974. Eventually, in the Amsterdam Declaration of 2001, more than a thousand delegates (including the world's four principal global change organisations) issued a statement that, in effect, endorsed him. It said: 'The Earth System behaves as *a single, self-regulating* system comprised of physical, chemical, biological and human components.'[8] This self-regulation – which takes place crucially within the narrow limits that the life forms undertaking it can tolerate – is the fulcrum of 'the balance of nature'.

When we heat limestone to make cement, or burn carbon-based fuels, we pour back into the atmosphere CO_2 that it took millions of years for 'Gaia' to lock up. Mostly we don't realise that this is happening because CO_2 is invisible. But imagine if every car driving down the street left a visible plume of smoke, and every plane left a vapour trail that took more than a decade to start fading noticeably. The Earth would be criss-crossed with the trails of our journeys, and that's actually how it is with CO_2 . . . except we can't see it.

Why does this matter? After all, CO_2 presently makes up less than one-twentieth of 1% of the composition of the atmosphere. It matters because the gas is a powerful modifier of the Earth's climate even in small quantities. Along with other greenhouse gases including water vapour, methane, nitrous oxide and ozone, CO_2 is molecularly 'tuned' to catch and re-emit the sun's warmth at a wavelength that warms the atmosphere. This energy would otherwise have been re-radiated off the Earth's surface and returned to outer space. But greenhouse gases keep it down to Earth. Increasing their proportion in the atmosphere is like wrapping a blanket round the Earth; it's like enclosing it in a gaseous 'greenhouse'. As the glass helps a greenhouse to warm up on a sunny day, so greenhouse gases have this effect on the whole Earth.

Human action is only one source of CO_2 to the atmosphere. Natural processes like plant decomposition and volcanic eruptions also produce massive amounts, though human output per annum is on average a hundred times that of volcanoes. From a combination of computer models, laboratory experiments and observations of what actually happens on the planet, most scientists are now convinced that anthropogenic greenhouse gas emissions are upsetting the atmosphere's balance and, with it, upsetting the balance of nature. Both the observed and the computer-modelled evidence concur in ways that strongly support theories of how global temperatures fluctuate with levels of atmospheric CO_2. Today these findings suggest that the stability of the Earth's climate is in serious jeopardy.

It is easy to blame science, or more accurately, the application of technology, for these concerns. But we must never forget that without first-rate science most of us would never be able to move beyond anecdotal speculation to realising that a problem actually exists. Consider the ozone hole that was such a worry to sun lovers a couple of decades ago. Most people would never have known what was contributing to their skin cancers were it not that elaborate scientific instrumentation was able to detect that which was invisible to the naked eye. The ozone hole problem was relatively easy to address. The effective implementation of an international convention led to changes in the types of gas used in refrigeration and aerosol cans. But the CO_2 problem is much more deeply structured into the economies and lifestyles of modern society than ozone gases ever were.

The evidence that CO_2 really is a problem comes from a variety of scientific approaches. One examines isotope balances in marine sediments. Another looks at the growth characteristics of fossilised plants. But perhaps the most convincing data comes from measuring the composition of minute air bubbles trapped in ice cores. These provide a cross section of time that documents changes over several hundred thousand years – double the time that human beings in our present highly evolved form as *Homo sapiens* have been on the planet.

In June 1999 a team of nineteen scientists from France, Russia and America published their study of the Vostok ice core in *Nature* – the most prestigious scientific journal in the world. This provided detailed information on a 3.6-kilometre-long ice core extracted from Vostok in Antarctica. It tracked cycles of atmospheric change over the past 420,000 years by analysing dust content and the composition of its minute air bubbles. The results strongly support the theory that there is a positive feedback relationship – one that amplifies itself – between the atmospheric concentration of greenhouse gases and the Earth's temperature.

The Vostok scientists observed that concentrations of CO_2 at the time when they undertook their study were 360 ppmv (parts per million by volume), and 1,700 ppbv (parts per billion by volume) for methane. The previous highest levels that they found in the historical record, as reflected in their core sample, had been 200 ppmv and 750 ppbv respectively. Their 1999 press release warned, 'Such levels are unprecedented during the past 420,000 years'.[9] Levels that were 10 or even 20 times higher than these have probably been present on Earth at certain times in the past half billion years. These would have been caused in particular by intense volcanic activity. It is true that the planet recovered from them, but not before life had suffered catastrophic upheavals such as the Permian mass extinction. It would, therefore, be unwise to use these times of high levels of atmospheric CO_2 to justify complacency today. The Earth at those times did not have to support the conditions that make possible our advanced civilisation.

Between 1999 when the Vostok study was published and January 2007, world CO_2 levels have risen further to 383 ppmv as measured by the US government's laboratory at Mauna Loa, Hawaii.[10] Presently they are increasing by more than 2 ppmv per annum and the rate of annual increase is growing exponentially. No informed person seriously disputes that this is caused by burning fossil fuels, forest destruction, intensive agriculture and cement making. Cement manufacture alone produces between 5% and 7% of anthropogenic CO_2 emissions. Not only does it take a huge quantity of (mostly) fossil fuels to run a cement kiln at 1,500°C, but the process of heating limestone itself gives off massive volumes of CO_2. Even the fabric of the buildings that most of us live in thereby contributes to global warming.

According to the government's Department of Environment, Food and Rural Affairs, in Britain today 42% of carbon emissions come from business, agriculture and the public sector, 26% from residential sources, 25% from domestic travel and 7% from air travel. The lattermost stirs the ire of environmentalists because its rate of increase threatens to undo efforts made towards energy conservation in other sectors. Also, much air travel is for reasons of status or pleasure. It highlights the tension between what we'd like to do and what we ought not do.

Since the approximate start of the Industrial Revolution in 1750, atmospheric CO_2 levels have risen by about 30%. Most of that increase has taken place since 1945. The worsening is now being driven by a combination of affluence in the West and the rapid industrialisation of countries like China and India. Industrial lifestyles are destabilising the planet. We stand in a cleft stick between expectations of ever increasing prosperity and limitations on planetary carrying capacity.

* * *

But while most scientists accept this analysis, powerful voices in public life seek to silence it. Martin Durkin was only the latest of a string of climate change 'deniers' but he set himself up to make a good case study. In *The Great Global Warming Swindle* Durkin denigrates much of the science that has just been discussed. He basically suggests that some 2,000 of the world's most eminent experts who have publicly put their reputations on the line by publishing or speaking out about climate change have got it wrong. As such, his views and those of his handful of interviewees might have been dismissed as scarcely worthy of debate. He is, after all, a film-maker and not a climatologist. But a great many people and vested interests wanted to believe the case he made. Although it feels like yielding to a distraction, that case needs to be looked at here. In a way, Durkin has done us all a service. He has provided a focus on the global warming sceptic position and this has pushed the other side to muster its evidence. As we have seen, the Royal Society already had concerns about climate change denial when it issued its website warning in 2006. Durkin's documentary crystallised a debate that needed to come into the open. He may have confused a willing public, but in so doing he has forced many people rapidly to become better informed than they might otherwise have bothered to remain.

Nobody on either side of the debate denies that global warming is a reality. Attempts were made in the early days to suggest that rising temperatures were merely an artefact of urban life. Many of the weather stations are in cities and the idea was put around that these have simply experienced a warmer microclimate due to energy released by buildings and transport. However, this theory failed to stand up when adjusted in the light of data from non-urban stations. Microclimatic effects do exist and are well understood, but global warming is pretty much across the board. Detailed records of world weather patterns exist from about 1850, when accurate thermometer recording began. Between 1906 and 2005 average global temperatures have risen by 0.74°C.[11] That may not sound like much, but it represents a massive increase in the energy that drives *weather* on the Earth's surface. Without doubt climate has fluctuated 'naturally' over time in the past. Evidence for this includes tree rings, coral growth, ice sheets and human historical records such as artists' paintings of the Thames freezing over, or medieval accounts of grapes flourishing in England. What distinguishes the change from about 1970 onwards is rate of increase. Presently temperatures are climbing by about 0.2 of a degree centigrade per decade. Most scientists consider that only a small part of this can be put down to 'natural' causes.[12] The bulk of it is almost certainly caused by us.

The most authoritative body in the world that looks at these trends is the Intergovernmental Panel on Climate Change (IPCC), set up by the United Nations Environment Programme and the World Meteorology Organisation. As we have seen, its November 2007 report says, 'Warming of the climate system is unequivocal'. Sea levels are rising, ice cover is shrinking and rainfall patterns are changing. The IPCC states that 'eleven of the last twelve years (1995–2006) rank among the 12 warmest years in the instrumental record of global surface temperature'. It concludes that the main cause of this is 'very likely' to have been anthropogenic – that is to say, our fault – and I am told by people who were present at the meetings where the 2007 report was finalised that the language would have been very much stronger had it not been for equivocation by such countries as America, Australia, China and Saudi Arabia. In a nutshell, such is the scientific consensus on which the Royal Society bases its concern.

* * *

But not so the school of Martin Durkin. Durkin maintains that the supposed link between CO_2 and temperature increase was stirred up by Margaret Thatcher as part of her vendetta against the coal miners. She wished to cut reliance on coal (i.e. the working classes) and oil (i.e. the Arabs), and to restore nuclear power to respectability (i.e. to the men in dark suits and white coats). Added to this was a twisted neo-imperial conspiracy to deny Africa its right to development.

To support his position, Durkin claims that NASA data refutes the idea that anthropogenic greenhouse gases drive global warming. The warming is real, he agrees, but the culprit is not the profligacy of consumer lifestyles. Rather, it is *sunspot activity*. Global warming in his view correlates closely with solar cycle data. As this originates on the surface of the sun, there's nothing we can do to stop it. We can all breathe a sigh of relief, curse the green fascists who scared us into guilt trips about our cheap foreign flights, and get back to sunning ourselves. Simple as that.

Writing in the *Daily Telegraph*, Christopher Booker reflected much of the anti-environmentalist comment that bloated the media following the broadcast of Durkin's film. He wrote:

> Only very rarely can a TV documentary be seen as a pivotal moment in a major political debate . . . Never before has there been such a devastatingly authoritative account of how the hysteria over global warming has parted company with reality . . . our own political establishment, led by Tony Blair and David Cameron, is lining up with the EU, the UN and that self-

promoting charlatan Al Gore. They propose measures that threaten not only to undermine the prosperity of the developed world but to rob billions of people across Africa and Asia of any chance to escape from the deprivation that kills millions every year.

Truly, this pseudo-religious madness has become by far the most important and all-pervasive political issue of our time.[13]

Within environmentalist circles Martin Durkin had already established sound credentials as a *bête noir*. Back in 1997 he had played a similar card in a TV documentary series called *Against Nature*. This, too, had portrayed green thinking as a conspiracy against the economic aspirations of ordinary urban and working-class people. George Monbiot did some investigative digging at that time and published his findings in his *Guardian* column. He headed it, 'The Revolution has been Televised':

. . . *Against Nature* was driven not by healthy scepticism but by shrill ideology.

If this were so, where might it have come from? At first we thought the Far Right might have been involved. But, over the last three weeks, another picture has begun to form. *Against Nature* IS the product of an extreme political ideology, but it comes from a rather different quarter: an obscure and cranky sect called the Revolutionary Communist Party.[14]

It would be only fair to say that when I checked out the website of the RCP, their current position does not seem to reflect that of Durkin – quite the contrary. One of their articles even affirms that 'respect for the dignity of Mother Earth and for the dignity of human beings go hand in hand'. It looks like either Durkin or the party have moved in different directions since 1997.

Durkin's axe to grind seems to be that sustainable development and 'green' thinking generally is a ploy by which the rich seek to deprive the poor of economic progress. He does have a point here. Conservation in the narrow sense of that word has often been the preserve of the foxhunting classes. It has gone hand in hand with conservativism, pheasant bagging and sucking rents out of servile tenants. I'm with Durkin that far. It is something that has driven my own work with land reform. Too often there's been a split between conservationists, who are passionate about wild nature, and social development, with its passion for people. But thankfully that divide has been closing. We need both conservation and appropriately sensitive development to provide sustained and dignified human livelihood in a rich and beautiful natural setting. What we don't need is cancerous over-development that promotes surplus rather than sufficiency, and results in exponential demands on the Earth's material natural resources.

But Martin Durkin seemed blind to these shifts that the sustainable development agenda has brought into synergy. The galloping narrative in *The Great Global Warming Swindle* told 2.5 million viewers – 11.5% of the audience share[15] – that it was telling 'the story about westerners invoking the threat of climatic disaster to hinder vital industrial progress in the developing world'. Durkin set environmentalism up as a 'straw man' – a caricature of his own invention that could be easily knocked down and ridiculed. In the end, however, it was his own programme that proved to be the straw man.[16]

Nailed by *The Independent,* Durkin admitted that his staff had changed a key graph purporting to present data from NASA. 'There was a fluff there,' he was forced to concede, adding: 'The original NASA data was very wiggly-lined and we wanted the simplest line we could find.'

The Independent put it rather more bluntly: '*The Great Global Warming Swindle* was based on graphs that were distorted, mislabelled or just plain wrong. The graphs were nevertheless used to attack the credibility and honesty of climate scientists.'[17]

Chris Merchant, lecturer in Earth Observation at Edinburgh University's School of Geosciences, made a further discovery that showed just how serious Durkin's fluffing was. He asked himself why it was, given that by this time it was 2007, Durkin's graph suggesting that rises in the Earth's temperature correlate with solar cycles had been truncated at 1980. He got hold of the raw data on sunspot activity and solar cycle length and used them to make his own graph. This revealed that during the quarter century following 1980, Durkin's supposed correlation radically broke down![18] It was, he said, 'the oldest trick in the book – selective use of data'. His findings were subsequently corroborated by a study published in July 2007 by the Rutherford-Appleton Laboratory and the World Radiation Center; as a member of the research team put it, 'You can't just ignore bits of data that you don't like.'[19]

Meanwhile, George Marshall of the Climate Outreach and Information Network drew attention to the fact that several of Durkin's experts were in bed with industry and political lobby groups dedicated to denying climate change. One of them, the Canadian specialist Patrick Moore, had credibility as a co-founder of Greenpeace. According to Marshall, 'since a very personal and painful falling out with Greenpeace in 1986 Moore has put his considerable campaigning energies into undermining environmentalists, especially his former friends and colleagues [for example] as lead consultant for the British Columbian Timber Products Association [and its effort to undermine] Greenpeace's international campaign to protect old growth forest there'.[20] Another of Durkin's interviewees, Fred Singer, was founder of The Science and Environment Policy Project which, says Marshall, 'aggressively contradicts climate science and has received direct funding from Exxon, Shell, Unocal and ARCO'. On 4 September 2006 Exxon, which incorporates Esso and Mobil, had been sent an unprecedented letter by the Royal Society. This accused the company of wilfully funding thirty-nine groups that 'misrepresented the science of climate change, by outright denial of the evidence that greenhouse gases are driving climate change, or by overstating the amount and significance of uncertainty in knowledge . . .'[21]

But the last word on Durkin's sources rests with Professor Carl Wunsch of The Massachusetts Institute of Technology – himself a Foreign Member of the Royal Society. Embarrassed by how Durkin's team had drawn him into their programme and used him out of context, he wrote: 'It never occurred to me that I was dealing with people who already had a reputation for distortion and exaggeration.' He added, 'I am angry because they completely misrepresented me . . . I am the one who has been swindled.'[22]

And Martin Durkin's response to all this? According to *The Guardian,* Durkin's production company, Wag TV, threatened to sue Professor Wunsch for defamation unless he made a statement denying that he had been misrepresented or misled.[23] On BBC Radio Scotland's *Riddoch Questions* programme, Durkin unrepentantly quipped that the CO_2 theory of global warming is nothing but 'a stupid quack theory'. Speaking to the online magazine *Spiked* about how he'd chosen his documentary's title, he said: 'I wanted to call it *Apocalypse My Arse*, but in the end we decided on *The Great Global Warming Swindle*. It's a provocative title, which helps with ratings.'[24] A key component of his motivation is thus laid bare.

When Durkin was emailed a polite list of questions by Dr Armand Leroi of Imperial College, London, back bounced the reply, 'You're a big daft cock.' A copy of this message went to Simon Singh, author of *Fermat's Last Theorem*. Durkin replied a little more fully to Singh's effort at mediation. Justifying his programme's climate change scepticism, he asked, 'Why have we not heard this in the hours and hours of shit programming on global warming shoved down our throats by the BBC? Never mind an irresponsible bit of film-making. Go and fuck yourself.'[25]

It is, of course, conceivable that Durkin is a scientific prophet who, like Galileo, has made discoveries that will revolutionise the way we understand the world. But more than likely there's another explanation. As Hans Eysenck – himself no great liberal – showed several decades ago, the dogmatic tendency to see the world in black-and-white tunnel vision is a characteristic of authoritarian thinkers whether of the political right or left. And as George Monbiot pointed out, it is hard to distinguish Durkin's supposed left-wing views on environmentalism from positions on the far right – for example, Paul Driessen in his book, *Eco-Imperialism: Green Power, Black Death*. The publisher's blurb for *Eco-Imperialism* claims, exactly like Durkin, that the green movement:

> imposes the views of mostly wealthy, comfortable Americans and Europeans on mostly poor, desperate Africans, Asians and Latin Americans . . . denying them economic opportunities . . . Many of its members are intensely eco-centric, and seem to believe that wildlife and ecological values are more important than human progress or even human life.[26]

Driessen's book goes on to attack corporate social responsibility measures that seek to raise social and environmental standards. Remarkably he likens them to the appeasement of Hitler. As he sees it, the 'unelected activist power brokers' of this latter-day fascism are such household names as The Nature Conservancy, Friends of the Earth, Amnesty International, the European Union and the United Nations![27] And Driessen's book, incidentally, is warmly endorsed by Durkin's disenchanted ex-Greenpeace chum, Patrick Moore.

It's a funny thing, but I've known only two people, both from the bastions of landed power, who've gone around publicly accusing social and environmental activists like myself of being Nazis. One was exposed by the courts and the other by a Sunday newspaper of actions or associations that suggested they themselves were the Nazi sympathisers!

We live in a strange world where millionaires who were once the scourges of the Earth like the late Sir James Goldsmith retire to become environmentalists, and some environmentalists, perhaps weary of never scraping much of a crust and with an impoverished retirement looming, become cynical lackeys to the corporate hardcore. Such is the fate of activists who campaign out of their egos rather than from some place deeper. They either burn out or sell out.

Chapter Two

BEYOND TIPPING POINT

The Great Global Warming Swindle debate had an interesting effect as I wrote the first draft of this book. Durkin's film was, on first viewing, disturbingly persuasive. It would have been great to have been able to believe it. I only found my intellectual bearings again after doing several days' research including talking with climate experts. The episode brought home how hard it is to be reasonably sure of what it is we think we know. I have more than one university degree and have been teaching human ecology at several universities for nearly two decades. And yet Durkin's spin managed to put me on a wobble at various levels.

It wasn't just his challenge as to whether or not climate change is anthropogenic. It was also his challenge to epistemology – the theory and structure of what passes for knowledge. I realised that just as Durkin had worked his charm by conjuring up impressive-sounding experts and weaving together soundbites from what they said, so many of my own opinions take shape in much the same way! The limitations of our intellects mean that it has to be like this. We all have to rely on experts. There's no way that I would be able to get my head around the complex mathematics that goes into building a climate-change model or the technology by which trace gas measurements are extracted from ancient ice cores. As such, trust in the work of others and a reasonable presumption of integrity is both necessary and precious.

In many societies, the trust held by the educated is considered a sacred responsibility. It holds together nothing less than the group's worldview. To abuse trust is to rupture the structure of psychological reality and open the doors to collective madness. That is why indigenous cultures, including those of the Celtic world, usually place the poet or shaman under a huge obligation to tell the truth. It may not be literal truth, but it does have to be poetic truth – metaphorical truth – which in many ways exacts a higher ethical demand. 'Take this for thy wages,' said the Queen of the Faeries to Thomas the Rhymer as she plucked for him the magic apple: 'It will give thee a tongue that can never lie.' That was why 'True Thomas' was able to 'rhyme'. It was his very commitment to truth that cleared the inner doors of creativity that would otherwise have stayed forever blocked.

In the modern world it is often the media that takes on this shamanic rôle. Producers like Durkin are the shamans that we trust to mediate truth by weaving story. That sacred trust is all the greater when audiences are numbered in millions. But equally, the temptation to play to the ratings is enhanced in such a competitive dynamic. The carry on with Durkin has at least pushed many of us to reflect on all this. At the end of the day we are each charged with maintaining the integrity of our own worldviews. We could all go down the road of becoming scientists who mathematically model climate change. We could all aspire to having a supercomputer out in the garden shed to test truth in ways that make compost out of numbers. The trouble is that to specialise in such ways would leave vast gaps elsewhere in our ability to assimilate and contextualise data meaningfully. In its more extreme forms it would generate the kind of social autism that is quite common in rarefied university departments – a hugely developed rational capacity but shadowed by what seems like a corresponding diminution of the ability to weigh up the values side of what matters. Sometimes when science claims to be 'values free' one wonders if it is not that the practitioners in question are 'values blind'!

In Scottish educational thought this problem is well recognised under a rubric known as 'the democratic intellect'.[1]

Democratic intellectualism accepts that knowledge will always tend to create elites. When such elites have power, the intensity of their specialisation gives rise to blind spots. A wise society must therefore 'democratise' knowledge. Specialist insight needs to be held within a 'generalist' framework that sets it in a bigger picture and tests it for blind spots and against the ethic of service to the community. For this reason, moral philosophy was once a cornerstone of the Scottish university degree alongside the natural philosophy of the hard sciences. Some of us would argue that the former should be restored to its pivotal place. We need such tools for thinking and to evaluate the kind of issues that Durkin's TV programme raises. We need, all of us, to be philosophers in the word's ancient sense of being a lover of the goddess of wisdom – *philo-Sophia*.

In trying to get to grips with complex issues I often find myself weighing up the people as much as the facts that they purport to present. In so doing I maybe fall short of the *Nullius in Verba* standard, but a generalist like myself just cannot claim to have a grasp on all the ins and outs behind the science of climate change. I do, however, respect the consensus of some 2,000 scientists whose work finds peer-reviewed summation in the various reports of a body like the IPCC. That is why, although many would say that their stance is too conservative, I will use their reports as my central reference point in this book. I therefore accept and acknowledge to my reader the constraint of taking much of what I read or hear on a basis of discerned trust within a framework of commonsense criticism, and to arrive at my positions partly by balancing up the opinions and gravity of different authorities. I can be content with that because I do believe that, as bodies like the Royal Society repeatedly demonstrate, science that is laid open to review by its own adequately informed peers has integrity. This is its humility and, therefore, part of its beauty. It is ultimately what protects knowledge from the arrogance of some of its holders.

The global scientific consensus has responded with impressive rapidity to new data on climate change as it exponentially piled in over the past two decades. I now find it fascinating to look back and recall that, when I first started to teach human ecology in the Faculty of Science and Engineering at Edinburgh University in 1990, it was a complete no-no to suggest that one 'believed' that climate change was taking place. In those days, sufficient compelling data had not yet been assembled, analysed, and published for scientific peers to criticise. Those of us who feared the accumulating environmental consequences of a fossil-fuel economy and who found ideas like James Lovelock's Gaia hypothesis persuasive had to be extremely careful how we phrased things to our students. Within the Centre for Human Ecology under the leadership of the inspirational Ulrich Loening, we agreed that it was wisest to talk only of 'possible climate change'. However, by the mid 1990s that consensus was in rapid transition. It became possible to start speaking of 'probable climate change'. And today, the question of possibility or probability is no longer at issue. Climate change is for real. The only uncertainty lies in the weighting between natural and anthropogenic attribution. Hardly anybody who has a reputation worth losing now disputes that we are in trouble, that that trouble is substantially of our own making, and that we're heading for more trouble.

* * *

I am not going to paint a detailed picture of climate change in this book. The IPCC's website carries in full the reports that do that with admirable clarity and accessibility for anybody with a high-school level of scientific education. A range of independent appraisals can also be found on the websites of the Royal Society, the Met Office, the Tyndall Centre and many other research institutions as well as government departments worldwide.

On top of these there is a growing wealth of carefully researched books for informed lay readers. These include, as we have seen, George Monbiot's *Heat: How to Stop the Planet Burning* (2006), Robert Henson's *The Rough Guide to Climate Change* (2006), James Lovelock's *The Revenge of Gaia* (2006), Fred Pearce's *With Speed and Violence* (2007) and *Six Degrees: Our Future on a Hotter Planet* by Mark Lynas (2008). One of the best books that I found late on in my own research has the same main title as this book (it is far from original, there being no copyright on titles), but aimed very much at the American market: *Hell and High Water: Global Warming – the Solution and the Politics* by Joseph Romm. Romm was formerly a Department of Energy advisor to President Clinton. I found his clarity akin to George Monbiot's, though many would find his prescription even more controversial. It includes cutting U.S. carbon emissions by two-thirds by taking such steps as building one million large wind turbines and '700 new large nuclear power plants while shutting down no old ones'.[2] Such are the disturbing implications of trying to move to a lower carbon economy while still maintaining business as usual for the American way of life!

For books taking the contrarian position on climate change, there's Nigel Calder's and Henrik Svensmark's *The Chilling Stars: A New Theory of Climate Change* (2007), Fred Singer's *Unstoppable Global Warming* (2007), and the old standby, Bjørn Lomborg's *The Sceptical Environmentalist* (2001), recently updated as *Cool It!: The Sceptical Environmentalist's Guide to Global Warming* (2007). At Christmas 2007 the latter was the top selling book under Amazon's 'global warming' category, which says something either about Lomborg's intellect or about what people want to hear him say. But what exactly is the scientific consensus about the state of the world's climate? What scenarios are we faced with? Here is a summary.

Temperatures

The Earth faces very serious warming within the lifespan of those being born today. Over the past century global average surface temperatures have increased by three quarters of one degree. The IPCC anticipates that they will rise by another 4°C and conceivably as much as 6.4°C by 2100 if world economies remain intensively fossil-fuel dependent. If radical remedial steps are taken to curb carbon emissions this could be kept down to 1.8°C, conceivably even 1.1°C, but warming of 0.2°C per decade over the next two decades is almost inevitable whatever happens because greenhouse gases persist for many years in the atmosphere and we keep on adding more. The IPCC is in no doubt that this is mainly attributable to human causes. Its February 2007 report on the physical science basis of climate change said:

> Most of the observed increase in global average temperatures since the mid twentieth century is *very likely* [my emphasis] due to the observed increase in anthropogenic greenhouse gas concentrations. This is an advance since the Third Assessment Report's conclusion that 'most of the observed warming over the last 50 years is *likely* [my emphasis] to have been due to the increase in greenhouse gas concentrations.' Discernible human influences now extend to other aspects of climate, including ocean warming, continental-average temperatures, temperature extremes and wind patterns.

When the IPCC uses the expression 'likely', it means an estimated probability of 66% to 90%. 'Very likely' means 90% to 99% probability or, as they otherwise express it, having a 'very high confidence' in such an assessment. Such care in the use of language and the cultivation of shared literacy around it has become increasingly important in scientific work. With climate change, we are dealing with heady concerns that need to be set in a well-ordered framework of probabilities. Failure so to do is one of the main difficulties in evaluating what's what, especially when the media scents the power of fear and uses it to amplify a single study so that it ends up punching above its weight. Let me show what I mean.

The European heat wave of summer 2003 is estimated to have killed 35,000 people, mostly the poor and elderly without air conditioning in their homes.[3] That figure seems to be reasonably founded as it is based on actual increases in mortality rates. So far so good. But a study attributed to the World Health Organisation claimed that heat-related epidemics are already killing 150,000 people a year worldwide. This statistic was widely reported in the world's media in 2005 and is regularly recycled by environmental organisations. However, it is an example of the kind of climate change science that, personally, I consider premature to employ in public awareness work. It was compiled mainly by one relatively minor official. He drew on a few narrow epidemiological studies – for example, research into Peruvian hospital admissions suggesting that a 1% increase in temperature leads to a 5% increase in diarrhoea infections. Such data was extrapolated up to a global scenario, creating what the official described as 'the best estimate we have'. But really, as one would trust that he if not the media was aware, until such an estimate is triangulated from a much wider evidence base and subjected to scrutiny in multiple peer-reviewed contexts, the 150,000 figure should be considered no more than a preliminary guesstimate.[4] Media hype perhaps arrests public attention, but in the end it indicts credibility.

Melting Glaciers and Rising Sea Levels

As ice caps and glaciers melt and warmed-up oceans expand, sea levels will inexorably rise. This is already happening. An Australian study published in *Geophysical Research Letters* in 2006 looked at records from tide gauges all over the world between 1870 and 2004. This found that average sea levels have risen over the period by 195 millimetres (about 8 inches). The average rate of rise was 1.44 millimetres a year. But it is accelerating. During the twentieth century the average rate was 1.7 millimetres a year, and since 1950 it has been 1.75 millimetres.[5] The 2001 IPCC forecast was that between 1990 and 2100, sea levels would rise between 90 millimetres and 880 millimetres. But data coming in from NASA and the UK's Hadley Centre now suggest that this is underestimated. Between 1990 and 2006 average global temperatures rose 0.33°C, and satellite data covering 1993 to 2006 reveals that sea levels have risen by an average of 3.3 millimetres a year. The IPCC had anticipated only 2 millimetres for this period.[6]

The probable reason why the IPCC's projections were too low is that the array of different models that they used incorporated ocean expansion, but failed to incorporate run-off from melting glaciers! This wasn't due to stupidity. It was the IPCC's extremely conservative process upon which its credibility relies. At the time, the science of glacial melt wasn't well enough established in peer reviewed journals to build it into the models. It is an example of the 'inevitable outdatedness' that was mentioned earlier.

NASA scientist James Hansen has repeatedly warned that as ice starts to melt, its albedo – the fraction of sunlight that it reflects back into space – reduces. This is because water makes ice 'black' and melting also exposes land, both of which soak up the sun's heat more effectively and thereby accelerate melting. Recent research suggests that many more of the world's glaciers are receding in contrast with the few that, owing to increased snowfall in some areas, are expanding. For example, China's news agency reported in December 2007 that glaciers in the west of the country, covering 20,000 square kilometres, have shrunk 7–18% in the past five years.[7] In addition, recent research suggests that when glacial meltwater sinks below the ice surface it lubricates and speeds up glacial flow into the sea. It also undermines the vast buttresses that, in some areas, support ice sheets over the sea. In particular, large parts of the West Antarctic ice sheet are supported, mushroom like, on top of narrow ridges of rock. If they suddenly collapse, sea level rise will speed up. One such collapse already happened in March 2002 when a chunk larger than the US state of Rhode Island fell into the sea.

Hansen himself warns that a sea level rise of 5 metres within the present century is within the bounds of possibility. In a paper published in 2007, he and his team add: 'Paleoclimate data show that the Earth's climate is remarkably sensitive to global forcings. Positive feedbacks predominate. This allows the entire planet to be whipsawed between climate states.'[8] Translated into plain language that means that the study of ancient climate change suggests that once a stable climate system gets messed about with, it can flip very quickly into another state – and sometimes at speeds measured scarcely in decades. Hanson's bottom line is that 'Civilization developed during a period of unusual climate stability, the Holocene, now almost

12,000 years in duration. That period is about to end.'

If temperatures rise by 4.5°C, which is within the IPCC bounds of possibility this century, the entire 2-mile thick Greenland icecap would start to melt, eventually causing sea levels to rise by 7 metres. A similar contribution would come from Antarctica, adding to the effects of ocean expansion. Such scenarios are the source of the more alarmist projections that we find in the media. For example, it can be pointed out that the last time the Earth was just 1°C warmer than it is today was 125,000 years ago during the Eemian interglacial period. Sea levels then were 7 metres higher than they are today.[9] How long this would take to come about, however, is a question shrouded more in speculation than in science. And that's part of our present-day problem tackling climate change. As we will see time and time again in this book, the long-term scenario looks very serious, but the rate of change is glacial in terms of the typical politician's duration of office or most people's willingness to sacrifice for future generations. For now, let us mark Hanson's point about Holocene stability having come about only in the past 12,000 years. We will return to that when we examine lessons from climate change in early antiquity.

Rising sea levels are a profound concern for low-lying countries like Bangladesh, Kiribati, and Tuvalu; for cities close to or below sea level like London and New Orleans; and for Holland with fully half its territory at or below sea level. It is already having impacts that make less of a splash in the headlines. The main way that I keep in touch with conservation thinking is through an outstanding little journal called *ECOS*, published by the British Association of Nature Conservationists. Its December 2007 issue carried an article by Alan Watson, Head of Land Use at the National Trust, who describes how 180 of the Trust's properties in England and Wales are at risk from flooding. Some of these are already subject to 'managed retreat'. He says, 'Climate change mitigation and adaptation are new imperatives for all our work and preserving the status quo is impossible.' The cruellest dilemma is that the very source of the National Trust's revenue further compounds the problem. This is because the Trust's carbon footprint 'is enormous, largely due to the extent of car-based visits to its properties'.[10]

Increased Rainfall and Freak Weather

Raising the Earth's temperature is like turning up the heat under a saucepan. It pumps more energy into weather systems. This makes them more turbulent, unpredictable, and it increases the rate of evaporation leading to more rainfall on average. A paper published by an international team of leading climatologists in *Nature* in 2007 confirms a trend towards increased rainfall in the twentieth century. As it says in technical jargon: 'Anthropogenic forcing has had a detectable influence on observed changes in average precipitation within latitudinal bands [and] these changes cannot be explained by internal climate variability or natural forcing.'[11] At the same time, the team point out that individual extreme weather events cannot categorically be attributed to climate change. Neither is it possible to predict with accuracy which regions will experience increased rainfall and which, paradoxically, will suffer drought because of changing wind patterns.

The general message is one of constant change being here to stay. Generally speaking the wet places will get wetter and dry ones drier. Tornadoes and storm surges are expected to affect areas not previously afflicted. Conservative estimates suggest that 80 million people worldwide will be at risk of flooding, 80% of them in the lower income countries of Asia. The dreadful injustice is that the people who are least to blame for anthropogenic global warming will most suffer the consequences.

Water Shortages, Food and Desertification

North Africa, the Middle East and the Indian subcontinent are expected to suffer from increased water shortages. Climate change is therefore expected to be a driver for regional conflicts and this will create refugee crises. A foretaste has already unfolded in Sudan. According to a 2007 United Nations Environmental Programme (UNEP) report, the civil war in Darfur can 'be considered a tragic example of the social breakdown that can result from ecological collapse'. Nearly half a million people are thought to have died.[12] A further 2 million are displaced, and 4 million are dependent on food aid. Declining rainfall since the 1930s has shifted the desert boundary 50–200 kilometres southwards. This puts a further 25% of Sudan's agricultural land at risk. As such, UNEP says, 'the impacts of climate change . . . threaten the Sudanese people's prospects for long-term peace'.[13]

It is difficult to comprehend the sheer scale to which we already make use of virtually every available corner of the planet. It has been estimated that human beings now appropriate 20–40% of photosynthetic product, and that this will possibly rise to 60% in the future.[14] In plain language, that means that we hunt, gather or harvest nearly half of what sunlight acting on green plants eventually produces. It is hard to imagine that when looking at semi-wild open landscapes, but even here, grazing generally keeps the grass pretty close-shaved from valley bottom to mountain summit. When pressed to such productive intensity ecosystems become more fragile. Soil quality degrades, loses its structure, and is easily washed or blown away. Some 16 million square kilometres of the Earth's surface suffers from water and wind erosion.[15] The area affected by 'tillage erosion' is presently unknown, but it is commonly said to take two bushels of topsoil to grow one of wheat. What's left when the soil has lost 'good heart' is often little more than a sandy rooting medium where agriculture becomes hydroponics in which crops rely entirely on the application of water-soluble fertilisers.

Yet again, the injustices are diabolical. Some of the most shocking statistics in the November 2007 IPCC report are buried in the small print. For example, it is anticipated that in some African countries during the present century, 'yields from rain-fed agriculture could be reduced by up to 50%'. But in

North America over the next few decades, 'moderate climate change is projected to increase aggregate yields of rain-fed agriculture by 5–20%'. Under threat of such 'punishment' is it any wonder that President Bush did so little to mitigate climate change?

New Pests and Diseases

Patterns of disease distribution will shift. By 2080 an additional 290 million people may be exposed to malaria. Warmer winters and more gentle frosts in the northern hemisphere mean that some insects will no longer have their numbers constrained by 'winter-kill'. This is already creating pest epidemics. One example is the spruce bark beetle which destroyed 10,000 square kilometres of forest in the Alaskan Kenai peninsula in recent years. Another is bluetongue, a disease of sheep and cattle. Bluetongue has moved steadily north from Africa and the Mediterranean. The first cases reached England in September 2007 and it continues a relentless northwards march, helped by the fact that fewer winter frosts encourage midges, the bite of which is the vector of infection.

As ecosystems become more stressed, vulnerable people will suffer a decline in their quality of life and this can be expected to impact on their capacity to resist disease. New diseases or opportunistic mutations of existing ones may become more common. The potential implications will be explored later in this book.

Ocean Productivity

Probably due to changes in ice cover, the abundance of shrimp-like krill that forms the base of food chains for fish and sea mammals in the Antarctic has declined five-fold from 1970s levels. Furthermore, as the oceans absorb some of the surplus CO_2 they become more acidic. Acidity has already increased by 0.1 of a pH unit over the past 200 years, which is around 100 times faster than has previously occurred naturally. Acid attacks shell and this injures many marine organisms.[16]

Once ocean temperatures rise much above 12°C their capacity to moderate world climate becomes compromised. This is because the water column becomes stratified as a layer of low density warm water forms on top of the high density colder water underneath. One can feel this when swimming in warm countries. It can be pleasant at the surface, but dive down and it rapidly cools even at night when there's no sun to heat up the top.

Like oil on water, the warm layer doesn't easily mix with the cold once the 12°C threshold is exceeded. This means that fewer nutrients get brought up from the bottom to the surface where the sunlight is. Plant life that relies on photosynthesis then no longer soaks up so much CO_2. The ability of the food chain to 'pump down' CO_2 in the shells and body fats of little critters into ocean sediments (and future rocks) is thereby compromised. This layering effect of warm water on cold is the reason why tropical seas generally look much clearer than temperate ones. They're less of a biological soup. It also explains the paradox of some of the richest fish stocks in the world being in the coldest waters. Food chains need nutrients and light more than they need warmth. In consequence, cold waters provide better feeding.

An added dynamic caused by the layering effect is that ocean algae and other micro-organisms give off a variety of chemicals. The most important is dimethyl sulphide (DMS) – sometimes called 'the smell of the sea'. According to James Lovelock, who is a recognised specialist in this field, DMS plays a key role in the formation of clouds over oceans. It provides chemical 'seeds' around which water vapour can condense into clouds. Puffy white cumulus clouds reflect sunlight away from the Earth. That's why they're always so bright to look down on from a plane. If less of them form over the oceans, because there is less DMS to seed them, water temperatures will rise even faster.[17]

Already global warming is having a noticeable impact on the mixture of fish species in many parts of the world. Scottish fishing boats now regularly find the odd Mediterranean species in their nets. A paper published in the *Journal of Marine*

Science in 2005 says: 'Based upon the observed responses of cod to temperature variability . . . stocks in the Celtic and Irish Seas are expected to disappear under predicted temperature changes by the year 2100, while those in the southern North Sea and Georges Bank will decline. Cod will likely spread northwards . . . and may even extend onto some of the continental shelves of the Arctic Ocean.'[18] That is an example of how climate change threatens even key cultural icons in our way of life. The great British fish supper may not disappear, but the most tasty versions of it might slip the net of our territorial control!

Economic Impacts

Perhaps the most influential environmental report that came out of 2006 was Sir Nicholas Stern's *Review of the Economics of Climate Change,* published by HM Treasury and the Cabinet Office. Although targeted at Britain, it has had worldwide impact, warning:

> Using the results from formal economic models, the Review estimates that if we don't act, the overall costs and risks of climate change will be equivalent to losing at least 5% of global GDP each year, now and forever. If a wider range of risks and impacts is taken into account, the estimates of damage could rise to 20% of GDP or more.
>
> In contrast, the costs of action – reducing greenhouse gas emissions to avoid the worst impacts of climate change – can be limited to around 1% of global GDP each year . . .
>
> There is still time to avoid the worst impacts of climate change, if we take strong action now.[19]

The truth is that nobody can really calculate the full costs of climate change. Stern's calculations are almost certainly a gross underestimate – a sugar coating to conceal the unpalatable taste of the pill that needs to be swallowed instead of just telling people the truth. Furthermore, most of the real costs can never be quantified in money terms – what price a human life, an entire species, or even an island nation or river delta where human settlement might be lost forever?[20] Nevertheless, Stern has done us a favour by at least starting to pose the challenges in terms that the conventional wisdom can kick around and count. It remains to be seen quite how this will sit with laissez faire economic interests constantly pushing to 'keep government off the back of business'. We are entering times when, more and more, those with power in the world have to be challenged: do they serve life, or just Mammon – the god of money?

Declining Biodiversity and Ecological Resilience

The IPCC's Fourth Assessment Report of November 2007 says 'there is medium confidence that approximately 20–30% of species assessed so far are likely to be at increased risk of extinction'. Note that this doesn't say that they will become extinct – it is 'only' an increased risk. But according to the distinguished biologist, E.O. Wilson, we are already losing some seventy-four species a day.[21] This compares with one a year that would be expected 'normally' during 'normal times' in evolutionary history, judging from what can be gleaned from the fossil record.

The faster habitats shrink, the more the rate of extinction accelerates because the interconnected web of life collapses. Like genocide, species extinction is a crime against all time. It is a loss of economic potential, beauty and each species' own intrinsic value – its sacredness – that can never be compensated. In northern Brazil and southern Africa higher temperatures and less rainfall are expected to cause massive forest loss. Forests that are currently net soaks of CO_2 due to photosynthesis storing carbon as wood and humus would then turn into net producers of CO_2 and methane as they decompose or burn. Models produced by the Hadley Centre suggest that the Amazon rainforest, the great lung of the Earth, could collapse – without a single tree being cut down – once the average global temperature rises by around 4°C. As we have seen, the IPCC considers that such an increase is the 'best estimate' of what is likely within a century if the world continues burning carbon on a business-as-usual basis.

One study published in *Nature* suggests that plant and animal species are currently moving 6 kilometres towards the poles every decade, and spring arrives 2.3 days earlier each decade.[22] It raises the consideration that an island like Britain has become a massive aircraft carrier slowly being towed towards the equator. It could seem quite attractive were it not for the increased storms and rain and the wider effects elsewhere. A bulletin from the Natural Environment Research Council states that 'in the North East Atlantic, the range limits of many zooplankton and fish have moved northwards by 500–1,000 kilometres'.[23] If this indicator also applies to terrestrial species, it somewhat adds jet thrust to our lumbering aircraft carrier.

Depending on the characteristics of the species in question, fish can usually swim from one latitude to another and therefore follow patterns of climate change seamlessly. But such mobility is no longer possible for many land-based plants and animals over heavily utilised parts of the Earth's surface. Plants that spread their seeds by wind and water – and especially the 'weedy' primary succession species or early-colonisers – are good at adapting. They specialise in seeking out unsettled conditions and disturbed habitats. But some of the 'climax' species that require mature stable conditions will be more and more in trouble. For example, a colony of rare butterflies might be happily ensconced in protected ancient woodland in Dorset. But if the specific plants on which their caterpillars feed are squeezed out by colonising species taking advantage of climate change, the butterflies may find themselves trapped with nowhere to go. Their route might be blocked by cities, farmers' fields, and uplands radically modified by drainage, intensive grazing and commercial forestry. In the absence of north–south 'wildlife corridors' of natural habitat to move along, the butterfly would be wiped out. Each time such local extinction happens another link is lost in the web of life. A few lost links can usually be withstood. This is known as 'resilience' in the ecosystem. It is why a rich web of biodiversity is 'a good thing'. But if many links drop out, the whole system becomes insupportable and can rapidly change. Forests can become grasslands and grasslands can become dust bowls and deserts. Climate change thereby threatens the entire planetary ecosystem. It will generally make it more 'brittle' – more easily broken – and harder to repair once the damage has been done.

Changes in Ocean Currents

The climate of maritime countries is substantially modified by ocean currents such as the Gulf Stream and El Niño. Britain lies at a higher latitude than Newfoundland. The north of Scotland shares the same latitude as southern Alaska. And yet our ports do not freeze in winter because the 'Gulf Stream' – technically the North Atlantic Current – wraps us around with a blanket of warm Caribbean water and its associated moist warm air.

The Gulf Stream is one the world's strongest ocean currents. It works like a conveyer belt. Warm water reaches the North Atlantic and its surface is cooled by icy Arctic winds. This causes it to contract, becoming denser. As some of the water freezes the surrounding water becomes denser still because salt is 'rejected' as the ice forms. This cold and salty water, being very dense, then sinks with a vengeance down to the ocean floor. It has to have somewhere to go, so it snakes its way back along the ocean floor roughly to where it came from in the first place. Such a 'pump' effect is called thermohaline (i.e. heat/salt) circulation.

But global warming potentially interferes with this process. As the icecaps melt, lower density fresh water mingles in to the system. This weakens the pumping effect. Some of the more sci-fi scenarios even warn of the Gulf Steam being 'shut off' with the paradoxical effect that global warming would lead to extreme cooling in the far northwest of Europe! However, most authorities consider this scenario to be alarmist. The IPCC thinks that major changes to the Gulf Stream are unlikely over the next century. We will explore this in greater detail shortly when I turn to Scotland as a case study.

Arctic Thawing

Fossils in ocean sediments collected from near the North Pole by ACEX – the Arctic Coring Expedition – reveal that ice has covered the Arctic for the past 15 million years. But 55 million years ago – during a period known as the Palaeocene–Eocene thermal maximum – the Arctic was sub-tropical. It had a climate like today's Brisbane or the Canary Islands. This was a short-lived episode of geological history, but it shows what can happen, even at the North Pole, if a sudden temperature increase of 5–10°C comes about.[24]

Dramatic temperature rise on this scale is thought to be caused by one or more 'tipping points' being reached. This is when a previously stable environmental system state reaches a point when the factors that regulate it can no longer hold together. As a poem written by W.B. Yeats at the end of the First World War puts it, 'Things fall apart; the centre cannot hold'.[25] Like a lorry tipping its load, the whole system slides into a completely different pattern of equilibrium. This is the greatest fear that climate change holds in store. The projections of the IPCC and the thousands of cautiously peer-reviewed studies it draws upon take a level-headed 'all else being equal' approach. Climate change 'alarmists' on the other hand, like NASA's James Hansen, simply ask not unreasonable 'what if' questions.

Since satellite monitoring began in 1978, the coverage of late summer Arctic sea-ice cover decreased in less than three decades by 30%.[26] But as we will see by the end of this book, even that rate of change is rapidly accelerating. A letter published in *Nature* in 1993 suggests that the world may already be on the slide. The Alaskan North Slope, previously a carbon sink in the form of peaty soils and forests, has now become a net source by venting more carbon to the atmosphere than photosynthesis captures. As peat soils rich in preserved vegetable matter thaw out, they dry out and decompose, releasing the greenhouse gases CO_2 and methane in the process. The writers of the *Nature* letter understatedly conclude 'that tundra ecosystems may exert a positive feedback on atmospheric carbon dioxide and greenhouse warming'.[27] But the consequences of such positive feedback would probably be unstoppable. The more the tundra thaws, the more carbon is released, and the warmer this makes the planet, thereby causing yet more thawing and so more carbon release.

The decomposition of peaty soils is only one of several potential tipping points. Another one is the so-called 'ice-albedo feedback effect' already mentioned, where melting ice becomes darker and absorbs more of the sun's warmth, thus accelerating the melting. Most worrying of all is the tipping point that was probably, for unknown reasons, the cause of the Palaeocene–Eocene thermal maximum. Deep in the sediments of the Arctic Circle vast quantities of methane is 'caged' in ice in the form of methane hydrates. According to the US Geological Survey, 'The worldwide amounts of carbon bound in gas hydrates is conservatively estimated to total twice the amount of carbon to be found in all known fossil fuels on Earth.'[28] As the Arctic warms up, these would start to release methane, and methane is some twenty times more potent as a greenhouse gas than CO_2 is: it degrades to CO_2 and water after about twelve years, while most CO_2 remains for as long as it takes for the Earth's biochemical processes to re-absorb it – perhaps for thousands of years depending on the levels of concentration.

Ironically, warmer temperatures in the Arctic are already hindering oil exploration on the North Slope in Alaska. At one time only the surface of the tundra thawed in summer. Now the melting is eating in to what used to be permafrost and what was once solid ground is turning boggy. This turns roads to quagmires, causing oil pipelines to sag, and buildings to subside.

* * *

The analysis just presented leaves us with a sobering thought. Within a couple of centuries it is not beyond the bounds of possibility that our descendents could end up squatting around a defrosted Arctic Ocean as guests mainly of Canada and Siberia. Already shipping companies anticipate that within two or three decades the mysterious Northwest Passage between the Atlantic and the Pacific will have opened and be a viable alternative to the Panama Canal – linking China and Japan to Europe and the eastern seaboard of America. And mark that time span. Such was the consensus in early 2007, but as we will see at the end of this book, it looks like having been too conservative an estimate.

Already the territorial stakes are up for grabs. In August 2007, news broke that two mini-submarines had planted a titanium capsule containing a Russian flag on the Arctic floor, 4,200 metres (two and a half miles) below the North Pole. A spokesman for the Arctic and Antarctic Institute in Russia told the RIA-Novosti news agency: 'This is a risky and heroic mission. It's a very important move for Russia to demonstrate its potential in the Arctic . . . like putting a flag on the Moon.'[29] The official Chinese news agency, Xinhua, put it in rather less gallant terms. It said, 'The gesture will symbolise Russia's claim to a large chunk of the Arctic shelf twice the area of Britain and estimated to contain up to 10 billion tons of oil and gas deposits, as well as vast reserves of diamonds and valuable metals such as gold, tin and platinum.'[30] One week later Canada responded with an announcement that it would build two military bases in the Arctic and commission six new naval vessels to patrol the Northwest Passage. In the words of its Prime Minister, Stephen Harper: 'The first principle of Arctic sovereignty is use it or lose it.'[31] Quite what the Inuit made of such geopolitical posturing was not reported.

Meanwhile, long-term investors smartly getting shot of their shares in Panamanian shipping might want to consider Antarctica as the hot real estate tip for the future. After all, there's money to be made from misery. And what better than 24/7 sunbathing in June and December, commuting seasonally between second homes by the North and South Poles!

* * *

I have summarised where the world, as a whole, appears to stand with respect to climate change. But each one of us must also ask the same questions locally. Different bioregions will vary, but the general principles of what we need to examine are similar over much of the world.

There's a great line in Shakespeare's *Macbeth* where Macduff asks, 'Stands Scotland where it did?' Ross answers him: 'Alas, poor country! Almost afraid to know itself.' That sums up where most of us are in terms of local climate scenarios. Here, then, is where Scotland would appear to stand. In many ways it is a set of scenarios similar to those for other temperate maritime nations of the northeast Atlantic.

Change is Already Here

The annual report on Scotland's Climate Change Programme, laid before ministers in March 2007, opens with the words: 'Scotland's climate has already started to change. Winter rainfall has increased dramatically, frost and snow are becoming rarer and the growing season has lengthened significantly. These changes are already affecting our plants, birds, fish, animals and insects. Our grass needs cutting in January. And we know that our climate will continue to change.'[32]

Rising Temperatures

By the 2080s – within the life expectancy of today's little children – average annual temperatures in Scotland are expected to rise (unless there is radical remedial action) by up to 3.5°C in summer and 2.5°C in winter, with relatively more warming in winter than in summer. Already, the autumn of 2006 was the warmest ever recorded with November of that year the wettest ever. Overall 2003–2006 were the warmest years since records began. These trends look set to continue in the future. Hours of sunshine are likely to decline over the next century due to increased cloud cover caused by a warmer Atlantic with higher rates of evaporation.[33] In short, Scotland can expect to get warmer but wetter.

Rain and Wind

Annual precipitation is likely to increase by 5–20% by the end of the century, with autumn and winter seeing the biggest increases. In contrast, spring precipitation will reduce, and there will be little change in summer. Already, as we saw earlier, longer term measurements show that Scotland has become much wetter since 1961. Average winter precipitation has increased by almost 60% in the north and west, and the annual average has increased by 20% for the country as a whole. Unlike more southerly parts of Britain, Scotland is unlikely to suffer from future water shortages.[34]

Raised intensity and seasonal concentration of rainfall will increase the risk of flooding. There may be an increase in the frequency of very severe gales, but a decrease in the number of gales overall. Violent storms will be more likely to cause 'storm surges' in coastal areas which, during spring tides, are likely to subject 26,000 properties to the risk of flooding. The 2006 report of the Scottish Environmental Protection Agency (SEPA) concludes: 'The weather will become more erratic and therefore less predictable, with a greater likelihood of extreme events.'[35]

Biodiversity

Some of Scotland's most distinctive landscapes – mountains, moors and coasts – could all see marked changes in the species compositions that they can support. Forestry is likely to be a net beneficiary of global warming. Carbon dioxide is a 'fertiliser' to plants, and a warmer wetter climate makes for a longer growing season. SEPA's 2006 report says, 'The flowering dates of some plants (particularly early spring species, such as blackthorn) have also changed and some butterflies such as the Scotch Argus have extended their range northwards while the southerly range of other species such as the Scottish Primrose is contracting.' In January 2008 the RSPB – the Royal Society for the Protection of Birds – issued a warning that the

Scottish crossbill, the only bird that is endemic to the UK – it is not found anywhere else in the world – faces extinction if the Scots pine forests on which it depends give way to a non-indigenous mix of species.[36]

Sea Levels

In line with the IPCC's global forecasts, the Scottish Government estimates a sea level rise of 'perhaps' 600 millimetres (about 2 feet) by 2080. This is probably an underestimate for the reasons that we have already examined. It means that some coastal settlements, agricultural land and communication links will be lost – especially if storm surges become more frequent. Special adaptive measures will be needed where, for example, nuclear power stations are close to the coast.

The Gulf Stream

Global warming is probably bad news for Scotland's skiing industry, but conceivably its fortunes could pick up again if the 'conveyor' of the North Atlantic Current – our extension of the Gulf Stream – slows down or even turns off. This turning off is thought to have happened previously. At the end of the last ice age an influx of fresh water from melting glaciers reduced the salinity of the North Atlantic. It caused Gulf Stream temperatures in northwest Europe to fall by 5°C in just a few decades. The irony was that as everywhere else was warming up, Scotland and surrounding areas became colder, thus locally postponing the end of the ice age.

Over the Atlantic as a whole, oceanic circulation may have slowed by nearly a third since 1992 – though care must be taken with this figure since not enough is known about the current's range of natural variations. At present it is not possible adequately to separate natural from human causes. The Natural Environment Research Council warns that 'a slowing or shutdown could happen over a few decades, and may bring

3–4°C lower winter temperatures to the UK and northwest Europe'.[37] James Lovelock has the following reflection on this:

> I can't help wondering if the climate of the British Isles and the Western part of northern Europe, which is now 8°C warmer than the same latitudes in other parts of the world, may be largely unchanged by global heating, because the 8°C lost when the Gulf Stream fails is just about equal to the predicted rise of temperature from global heating. Perhaps this is no more than wishful thinking, and we will certainly have to pay through the loss of land as the ocean rises to repossess it.[38]

Land Use and Landscape

One of the spin-offs of climate change is already visibly affecting Scottish landscapes. Windfarms have split the green movement down the middle and left many of its adherents feeling bitterly divided within themselves. A windfarm is elegant in its own way, but huge ones – turbines of over 100 metres have become the norm – impose a mechanical dynamic on a landscape that constantly steals the eye's attention. They diminish the mind's ability to settle and to find rest. If people are to be motivated to live more sustainably with the Earth, they need to be able to connect closely and not have to turn away, either to look or go elsewhere. This creates a terrible dilemma, one reminiscent of the Vietnam war maxim: 'We had to destroy the village in order to save it.'

My view is that windfarms, hydro and other renewables such as wave power are a vital part of the energy mix needed for Scotland's future, but location, scale and local sense of ownership are everything. It is criminal that the planning framework surrounding these has so far been very weak. Offshore is the place for very large wind turbines. On land, I agree with those environmental organisations that argue that nothing taller than 50 metres to the vertical blade tip should normally be approved and that a criteria-based planning approach should be implemented to govern appropriate zoning.[39] Personally I would not allow any private landowner to take the initiative and be the primary beneficiary of a development that is going to impinge upon the entire surrounding community. Instead, a differential planning system should be introduced where community-backed proposals are given favoured consideration. Linked in to Scotland's existing land reform programme, this would encourage the movement of land out of private hands, the price of the sale underwritten by some of the capitalised proceeds of future energy generation.

To me, and I think to many Scots, there is a huge psychological difference between a renewable energy development that benefits local communities and one that merely enriches absentee landlords and the corporations that they bring in. A typical modestly sized wind turbine in Scotland presently generates around £50,000 a year assuming it can be connected to the national grid. This both reduces the local carbon footprint and puts funding for community regeneration onto a self-starting and sustainable footing. For example, the community land trust on the Isle of Gigha is now substantially supported out of the proceeds from their three wind turbines – the so-called 'Dancing Ladies' of Faith, Hope and Charity – which provide the island with some £150,000 a year.[40]

It is true that these small-scale initiatives contribute only marginally to the needs of city dwellers. On their own they are not enough to provide renewable energy for the nation as it is presently structured. But that structuring is at the heart of the problem. The UK electricity market is expected to expand by 55% by 2050,[41] so the argument that beautiful places should be sacrificed en masse to provide a 'sustainable' fig leaf against the backdrop of a licentious economy is a little galling. We need to balance how we produce energy with the equally big but much less comfortable question – why do we use so much?

Housing

Buildings account for about one third of energy consumption and, in the UK, they contribute 27% of CO_2 emissions.[42]

They are presently amongst our least efficient energy uses. To remedy this, Scotland's housing stock should be improved to reduce the massive losses caused mainly by poor insulation. Roofs should be used to gather energy, especially for water heating. The Scottish Ecological Design Association concede on their website, 'We see ecological design as a discipline which is still in its infancy.' There is still a long way to go in making it affordable. But while ecological housing is generally expensive at present, it is cheap once up and running and therefore makes a good investment. The materials and designs used can vary greatly, but where energy is concerned there are two key principles. Minimise embodied energy – that which is embedded in the manufacture of construction materials – and maximise energy efficiency during a building's lifespan.

In hot climates this prescription means an emphasis on cooling and water conservation. In the Scottish climate it means three things: insulation, insulation and more insulation. This is because water and space heating consume 84% of the energy used in an average British house.[43] All other uses – for example, domestic electrical appliances, lighting and cooking – account for only 2–3% each. The single most effective steps that most householders can take to help save the planet is to insulate the loft, walls and windows, install a modern boiler, slip a jacket round the hot water tank, turn down thermostats and wear warmer clothes, turn off appliances on standby, have an efficient fridge and freezer, and replace the tumble dryer with a clothes line. It may sound boring, but it's the only way to live within our energy means.

There is much speculation about 'zero carbon homes' defined as a building with 'zero net emissions of carbon dioxide from all energy use in the home'. In practice, this is only feasible where there is space to gather energy and when domestically generated surplus energy, for example from solar panels on the roof or a wind turbine in the garden, can be fed back into the National Grid to offset times when nature doesn't provide.

At present solar voltaic energy is very expensive to produce. But the idea of having every roof tiled with panels that passively feed sunshine into the national grid is immensely appealing. I have a German friend who produces more electricity than his house consumes in this way. He expects to repay his installation loan within ten years because the German government forces his supply company to buy back electricity at three times the average price that they sell it for. Superficially that sounds terribly unfair on the utility, until one realises that in Germany the spot market for electricity at midday is six times the average price. And that's precisely the time of day when the sun shines most strongly and solar panels on the roof are most likely to be running in surplus. As such, solar voltaic panels may be a very expensive way to generate electricity, but their visual impact is minimal and the energy they produce is particularly high grade in terms of its usefulness. It lends partial justification to the maxim 'The future is solar'.

Energy and Transport

Energy costs will rise as strategies are adopted to reduce carbon emissions and cheap sources of energy are used up. Higher fuel prices and carbon taxing threaten the transport sector and, therefore, economic activity closely linked to it, including prices in the shops. Much of Scotland's present prosperity, especially in more remote rural areas, is contingent on fossil-fuel based transport. Measures to combat global warming may encourage more people to take their holidays at home, but if car, air and travel by ship are curtailed, living in a remote area like the Highlands and Islands will become proportionately more costly. Offsetting these costs by renewable energy production is one reason why there has been strong support in some local quarters for what would have been Europe's largest windfarm on the Isle of Lewis, even though nearly 90% of residents reportedly opposed the scheme. Indeed, the strength of that opposition was quite remarkable and demonstrates the political sensitivity that shrouds many aspects of the energy debate. By February 2008, when the Scottish Government announced that it was minded to issue a rejection of planning permission for the Lewis windfarm, a stunning 9,859 letters of objection had been received (with a clear majority from locals) versus a paltry 77 in favour. Meanwhile, the island's Labour MSP had been unseated owing to his pro-windfarm stance. This resulted in a challenge to the overall face of the British constitutional settlement because it allowed the Scottish National Party to gain slender control of the devolved Scottish parliament.

Whichever way one looks at it, living a modern way of life in a remote area is energy intensive. Even travel by boat has a high carbon footprint because of what it takes to thrust perhaps several tonnes of metal per passenger through the water. The only way to really cut our personal carbon quotas is to start living more simply. As we will see later, this is not possible in all walks of life, but some people do partly demonstrate it in the choices they have made. For example, near Ullapool on the remote Scoraig peninsula, a number of families have followed a relatively simple way of life for the past four decades. They have no road. Some of them heat their homes largely with home-grown wood – it takes about 12 acres (5 hectares) per household when sustainably harvested. All have their own mini windmills and solar panels, and most rarely go to the city for shopping. But while this way of life has enriched their children as human beings, it does challenge many of them as teenagers. It is hard for young people to embrace a lifestyle that their school friends may denigrate as 'hippie', eccentric, or just plain poor. It is even harder to convince people in the mainstream that such pioneer lifestyles might contain the seeds of both present and future right livelihood!

Back in the 1960s, when I was growing up on Lewis, there were maybe only three cars in the average village. There would be five buses a day and for most people that was sufficient. In between buses you'd hitchhike. Today it has all changed. A typical contemporary family home will have three cars parked outside. Most people no longer live by the bus timetables. Hardly anybody picks up hitchhikers – you just don't know what you might be getting, and in any case, you're going so fast that you've passed before you see them. It is our very prosperity that drives our energy profligacy. I asked an old woman in the village of Gravir if she still cut peats for fuel. She pointed to her plastic central heating oil storage tank. 'That's the peat stack these days – the green peat stack!' she laughed. We're all a part of this and we benefit from it. It's part of our material prosperity. And yet, something of the human is diminished. Who knows, if the imperatives of climate change force us to 'go back', we might just find ourselves inching forward.

Chapter Three

DEVIL'S DILEMMAS

A Canadian friend told me of the time a First Nations chief invited him home. The chief was an environmentalist, respected for inspiring white folk's conferences with speeches about the coloniser's degenerate way of life and the imperative of restoring traditional values.

My friend footed his way out to the reservation, only to be gutted on finding the chief and all his family driving around in clunky gas-guzzling sports utility vehicles. 'How come this squares with your environmentalism?' he asked. 'Shouldn't we be thinking of the planet and conserving oil?'

'It squares no problem,' quipped the chief. 'The quicker we use all that oil up, the sooner we'll have to find the solutions.'

Well, he had wit! But on a small scale this argument connects to a much more influential one called 'Peak Oil'.[1] It's a scenario that's been around since at least the 1950s, summed up in the Saudi saying, 'My father rode a camel. I drive a car. My son flies a jet airplane. His son will ride a camel.' The point is that the world has almost reached, or perhaps just passed, the peak rate at which oil can be extracted. We've used up half the known reserves and from now on, we're increasingly running on empty.

Some people think that Peak Oil is quickly going to knock our CO_2 profligacy on the head. Unfortunately I'm not so convinced. From a global warming perspective, I even see Peak Oil as a distraction. There are still vast areas in Saudi Arabia, Siberia and, heaven forbid, in the Arctic and Antarctic, that are virtually unexplored or currently protected from exploration, but with high prospecting potential. And we don't even need to go down that road. Peak Oil arguments generally look at reserves within current economic frameworks. Yet even at present prices, oil remains cheap. In America in mid 2007 petrol or 'gasoline' retailed at around $3 per US gallon. That's 40 pence a litre, which is less than the forecourt price of most brands of bottled water. Admittedly the water is not sold in pumped volume, but it remains a salutary thought. Americans will say that a dramatic oil price rise would devastate the economy, but that hasn't happened in Britain where prices at the pump are two and a half times the American ones because three quarters of what we pay is tax.[2]

When it comes to air travel, the economists at IATA, the International Air Transport Association, tell me that fuel presently comprises about 25% of global airline operating costs. If aviation fuel costs doubled, the price of a £100 airline ticket, if all other factors remained equal, would rise only to £125 – an increase not of 100% but of 25%. In reality, other costs would be affected by the oil price hike and so the rise in ticket price would be a little more than I've just stated. But the general principle stands. Energy remains dirt cheap and prices would have to rise considerably to curb demand.

The seriously bad news from a climate change perspective is not that the world is short of carbon and hydrocarbons, but that we're slopping in them. The Earth's geological processes used ancient sunlight to lock excess carbon up and out of harm's way in forms that include such combustibles as mineral oil, gas, coal, oil shale and tar sands. The tar sands of Athabasca in Canada alone contain enough oil to boost known world reserves by over 50%. At one time these were considered commercially unviable, but new processes and higher prices have changed all that. The process is filthy and inefficient. It's staunchest critics, such as the campaigning magazine *The Ecologist*, assess it as being the least efficient of all energy sources in terms of its carbon footprint.[3] The trouble is that the higher oil prices rise, the more such inefficiency pays dividends. The bottom line is that with oil prices rising from $51 to $95 during 2007 and breaking through the $100 ceiling early in 2008, the industry now finds tar sands profitable. Try putting 'Athabasca Tar' into Google, and one of the first websites to come up is an environmental one dedicated to 'the worst places in the world'! Athabasca wins the booby prize for the landscape devastation and oil contamination that it causes. The response of companies like Shell is to implement what they call 'a good neighbour policy in order to develop a mutually prosperous, long-term partnership', particularly with affected First Nation peoples.[4] I have not researched this enough to know how that policy goes down locally, but globally it begs the old questions: 'Who is your neighbour?' and 'What are the power relations between you?'

In July 2007 the National Petroleum Council in the US published a report forecasting global oil and gas demand through to 2030. It acknowledged that 'over the next 25 years, the United States and the world face hard truths about the global energy future'. World energy demand has increased by 60% over the past quarter century and the Council forecasts a further 60% increase on present levels in the next quarter century. Nevertheless, they say they're ready for the challenge. The report claims: 'The world is not running out of energy resources, but there are accumulating risks [and] to mitigate these risks, expansion of all economic energy sources will be required, including coal, nuclear, biomass, other renewables, and unconventional oil and natural gas.'[5]

The problem in terms of carbon emissions is that elasticity of oil price reflects in elasticity of supply. The more the price stretches, the more that ugly alternatives get pulled into becoming viable. Tar sands and oil shales are only one 'unconventional' source. Another is coal. China and South Africa have enough coal reserves to last for hundreds of years at their present rates of consumption. Even at the exponentially increasing rates of use now emerging, their reserves should last at least another fifty years[6] – enough for the present generation of decision makers to see when the locomotive's going to run out of steam and say, with a relieved sigh, NIMTO – 'not in my term of office'! The reason why coal is such a potential environmental threat is that it can easily be converted into oil. During the years of the apartheid sanctions, South Africa couldn't buy oil on international markets. Scientists at the state oil company, Sasol, developed technologies first used by the Nazis that converted coal into 'synfuels' – synthetic liquid fuels. As with tar sand and shale extraction, it is a filthy process. It churns out both CO_2 and sulphur pollution. But for South Africa, synfuels today remain a cornerstone of the national energy policy. The Secunda plant – presently the only one of its type in the world – has the capacity to produce 160,000 barrels of transportation fuel per day from coal. This provides 30% of South Africa's transport requirements. Such is the success that Sasol are now trying to export the technology for what their website describes as 'a cleaner and more reliable energy supply'!

In short, I would have loved to have shared our First Nation chief's Peak Oil optimism and let rip participating in his guilt-relieving opulence. But while nobody doubts that Peak Oil must hit home at some time, I don't see it happening fast enough to put the brakes on global warming. Once 'gas' on the American forecourt starts to cost considerably more than the bottled water, we may be surprised at what efforts can be mustered to find more. We may be disturbed at just how much more carbon and hydrocarbon still remains to be scraped, drilled and blasted from out of the belly of Mother Earth.

* * *

What, then, ought we to do in practice to avert climate change? I must remind my reader that this book is not intended to be a technical manual of mitigation options. I am neither qualified nor desirous of repeating the detailed analyses and debates published elsewhere. My interest lies in getting to the psychological and spiritual determinants that underpin the scope for manoeuvre. But to get there, we need at least a rough map of how the conventional ground lies. Here, then, are some key pointers to the sorts of action that are necessary for us in the West to consider.

Energy Taxation

We must introduce sweeping energy taxes, calibrated so that heavy consumers – the rich – bear the consequences of their consumption. These taxes – many of which already exist but in forms that are too easy-going – will need to be on transport (especially flying), on fuel (especially for vehicles), and on domestic and industrial energy uses. Tax structures must be based not on the cost of getting energy out of the ground or its cost on global markets, but on the contribution that any given energy source makes to global warming. This will favour the development of energy efficient technologies and renewable sources which, at present, may seem prohibitively expensive only because the fossil fuels they compare with are so obscenely cheap and, even, state subsidised.

Shifting the base of taxation away from people's labour and profit and onto energy simultaneously tackles both the energy that we daily consume and that which is 'embodied' in the manufacture of goods and services. By definition, many luxury goods embody high levels of energy. Gold and diamonds are expensive partly because vast amounts of energy and energy-intensive infrastructure are required to crush huge volumes of rock in getting them out of the ground. This renders energy tax a relatively 'progressive' tax – one that takes from the rich more than the poor. Intelligently designed tax structures could protect the fundamental needs of everyone, especially the poor. For example, a progressive scale of charge thresholds built in to domestic electricity and gas meters could protect the poor and the elderly from going cold while suitably penalising heavy users.

Energy tax proceeds could similarly be used to subsidise efficient forms of public transport. At present it typically costs three times as much to travel to a continental European destination from Britain by rail as it does by air. This is partly because international agreements set in place to subsidise the airline industry prevent aviation fuels from being taxed. It's not just the whisky that's duty free at airports! However, if taxation was used to turn things round to the point where air travel cost three times the rail fare – in the way it used to do not so many years ago – then things would be very different. In an encouraging move the budget airline, easyJet, has broken ranks with the rest of the industry and suggested that the present air passenger duty should be replaced with a tax based on the greenhouse gas emissions per flight.[7] For easyJet this might be easyTax since many of their flights are short hauls. They might therefore gain from by not having a flat rate of duty. Nonetheless, it is encouraging to see an industry leader starting to talk about a form of fiscal intervention that was, until recently, off the Richter scale of thinkability.

Carbon Emission Quotas

A measure that is related to carbon taxing is the tradable carbon emission quota. Here the idea is that a nation's fair limit of carbon emissions is divided up amongst energy users – both industrial ones and, potentially, individuals – and a market is then allowed to develop to balance supply and demand. Known as 'cap and trade', this approach theoretically means that the rich would have to buy energy quota from the poor, thus providing another way of integrating social justice with environmental sustainability. Taken to its extremes, we could have to pay for goods and services in two currencies – in normal money, and in an emissions cost currency.

The European Union presently has the world's only mandatory emissions trading scheme. Started in 2005, it is a central pillar of the EU's collective Kyoto Protocol target. Brussels issues permits to big CO_2 polluters based on an assessment of their 'need to pollute'. Corporations then trade quotas amongst themselves according to their carbon surpluses or deficits. However, this sort of 'free market environmentalism' or 'green capitalism' has earned much criticism. Because the quotas are a hand-out it means that the existing status quo has been handed a market asset. This may have helped industry to accept the scheme without too much pain but, with quotas having been over-issued and therefore trading very cheaply during the first phase of the programme, there is little evidence that they have created a real incentive to invest in driving down emissions. An alternative approach would have been to rent out the quotas, thus raising revenue as a proxy green tax. That would have been too effective, and so industry would have put up a harder fight. At the end of the day I find quotas an irritation. One has to ask why bother creating an entire new service sector of energy accountants and auditors when a simple energy tax – levied at the oilwell, mine head or port of entry into a country – would be difficult to evade and would provide an incontrovertible incentive to cost carbon pollution into all levels of the economy.

Renewable Energies

I am not going to review the debate here about the potential or lack of potential for various renewable options to resolve global warming. There are plenty of other books that do this. The uncomfortable bottom line of them all, as George Monbiot's *Heat* shows with disturbing clarity, is that the sums only add up if we also choose to live modestly: unless our demand for energy is heavily reduced through increased efficiency, legislation and collective lifestyle change, then the contribution of renewables can only be a small slice of the cake.

There are two general aspects to the renewables debate. One is large-scale generation to provide for the wholesale needs of an industrial society. The other is micro-generation linked to people personally choosing relatively self-sufficient lifestyles, perhaps in the hope that society as a whole might one day emulate them. Let me look briefly at the first here, and at the second in the subsection that follows. And let's start by taking Denmark as an example – a country widely viewed as the world leader in renewables.

I find that the statistics on Denmark's achievements that appear in the mass media and green publications are radically diverse, but primary data from the Danish Energy Authority's *Energy Statistics 2006* shows the following. Since 1980, Denmark's gross domestic product has increased by 75% while energy consumption has gone up by only 6%. According to the Authority, 'this means that each unit of GDP demanded 40% less energy in 2006 than in 1980'.[8] So far, all very impressive.

In 2006, the contribution of renewable energy to Denmark's domestic electricity demand was 26.5%. But electricity is only one aspect of energy consumption. Transportation fuel and coal for power stations are amongst the others. When renewables are expressed as a share of Denmark's overall energy consumption, their contribution is brought down to a rather less impressive 15.6% of the total. The rump of Denmark's energy – the remaining 84.4% of it – still comes from greenhouse-gas polluting coal, oil and gas. Indeed, the amount of electricity that Denmark produces from renewables is approximately equal to what it gains from North Sea gas.

Denmark's bottom line achievement remains considerable. Taking together renewables, the switch to gas, and efficiency gains in both energy production and consumption, the country's CO_2 emissions fell by 13.6% between 1990 and 2006. Total emissions declined across all major economic sectors – agriculture, industry and households – with the significant exception of transport, which has increased its carbon footprint by 22.5% since 1990 (again reflecting that sector's devastating capacity to undermine achievements made elsewhere).

The downside to Denmark's success is that they've now done the easy bit. Short of major technological breakthroughs, the further they try to push the limits the more they will hit up against diminishing marginal returns. For example, achieving the next 13.6% cut in CO_2 emissions is unlikely to be helped by finding further fields of North Sea gas. The substitution of gas for coal has been an easy move for European nations striving to impress with their Kyoto emissions targets. To burn coal produces 77% more CO_2 per unit of energy than natural gas.

Oil emits 40% more than gas.[9] Substitution is therefore a relatively painless solution, but it only helps while finite gas reserves last. The bottom line is that an enlightened country like Denmark has led the way in cutting greenhouse gas emissions, but it remains a major contributor per head of population to global warming.

In contrast with Denmark, Britain and Ireland lie at the positively ugly end of the European carbon league. We get just 1–2% of our energy from renewables and little effort has so far been made to boost home energy efficiency. Most of our Kyoto obligation has been achieved by the sleight of hand we have just examined – juggling gas over coal. A Tyndall Centre report for Friends of the Earth and the Co-operative Bank puts the whole thing in rigorous perspective. Of the UK government's present targets it says:

> It is an act either of negligence or irresponsibility for policymakers continually to refer to a 2050 target as the key driver in addressing climate change. The real challenge we face is in making the radical shift onto a low-carbon pathway by 2010–12, and thereafter driving down carbon intensity at an unprecedented 9% per annum, for up to two decades.[10]

Alternative Livelihood

Some of the people I most admire in life are folks living very simply close to the land, getting most of their needs locally. It means accepting a level of demand reduction that, by most people's standards, would comprise poverty were it not for their evident contentment and the inner vibrancy that comes from taking hands-on control of one's own basic material needs. That said, even a place like the example already mentioned at Scoraig still depends heavily on the outside world for things like medical and educational services, and the capital that incoming residents bring when doing up houses.[11]

Intentional (because they are founded on 'intent' rather than accident or necessity) alternative communities exist all over

Europe. Some, like Le Viel Audon in the French Ardèche closely mirror traditional ways of life. At their Ardelaine woollen mill they have rejuvenated the region's sheep industry around restored water-powered machinery. Others, like the controversial 'freetown' of Christiania in Copenhagen, attempt to live more sustainably in an urban context. Their mission statement says:

> The objective of Christiania is to create a self-governing society whereby each and every individual holds themselves responsible over the wellbeing of the entire community. Our society is to be economically self-sustaining and, as such, our aspiration is to be steadfast in our conviction that psychological and physical destitution can be averted.[12]

These pioneers demonstrate that preliminary steps towards the green idyll are possible. And yet, at the end of the day, one of the ironies of alternative living is that it tends to highlight just how much we are all still connected in to 'the system'. We can reduce our dependence on the mainstream, but it is not possible entirely to 'drop out'. Partly this is because the system has stitched up the structures of capital, land ownership and planning, and not even a complete anarchist can stay above the law and property constraints for long. But it is also hard to sustain a completely separatist way of life because we are all interdependent, and parts of what the mainstream offers can be life-giving. For example, when something goes wrong with that intended natural childbirth out in the woods, the appearance of an ambulance's flashing blue lights over the brow of the hill is a relief! And when old age comes not alone, the support structures of the state can be gratefully received.

I have noticed, too, that with rare exceptions, those who make the biggest noise about having 'kicked the system' tend not to be the most functional members of their groups. That's a 'tough love' truth to name, but 'rebels without a clue' aren't going to change the world in a hurry, and too many alternative experiments (they are rarely referred to as more than 'experiments') have failed because they've become bogged down as magnets for the axiom that 'God gets what man rejects'. That is to say, alternative lifestyles attract a high proportion of social casualties who, perhaps through no fault of their own, can't make it elsewhere, but who freeload and impose a psychological burden on what they take to be more accommodating structures. As any person living a traditional way of life close to the soil will testify, simple living takes constant hard work and well-grounded relationships. It's not cut out for those who merely want to rough it on benefits for a couple of summers, or the hippie with an inheritance who buys their way into smallholding and, after planting a few nut trees or putting a hi-tech renewable energy gizmo on the roof, pretends they've achieved 'permaculture'.

These foibles are some of the downsides of the alternative movement. The upside is that its experiments are signs that hint that a lower-impact and more wholesome way of life can be a possibility. It can be a re-tribalisation – a re-building of extended family – that meets fundamental human emotional and economic needs grounded in a community of place. In terms of reduced carbon footprint this involves a loosening of connections with consumerism, growing part of one's own food, creating one's own community entertainment, minimal use of cars, recycling from what others throw away, and an array of micro-energy solutions including wind, solar, micro-hydro, ground source heat pumps, combined heat and power, wood burning stoves, woolly jumpers and the best ways of all to keep warm – namely, hard physical work during the day and gentle human affection at night! All this needs to be better recognised as a public good instead of being seen as a bit of a joke by a mainstream that doesn't quite know what to make of it.

To strengthen options for alternative living, planning restrictions should be eased to assist the creation of low-impact settlements designed on ecological principles. Better public understanding needs to be created as to why community and re-localisation must be at the heart of sustainable lifestyles. It is only if we can find fulfilment in close proximity to one another and local place that we can hope to stop sucking what we need from all over the world. As such, there is much that can be criticised in alternative living, and much to be found ridiculous in some of its more wacky practitioners: but not as much as we take for granted in the planet-wrecking rat race.

Energy Demand Obviation, Buildings and Industrial Ecology

There is massive scope to use energy more efficiently thereby potentially reducing overall demand – what is known in the trade as 'energy demand obviation'. This can apply to any energy sector, but let us take the built environment as an example. One third of total energy use is applied to buildings – including that which is embodied in their construction, consumed in their heating or cooling, and embodied in maintenance and servicing. Making this more efficient requires both innovation and legislation to encourage sustainable building solutions. It means supporting the construction industry to create new products, introducing green building regulations, and opening up new markets for innovative construction products in line with evolving environmental awareness. The principle behind energy obviation is that, wherever possible, needs should be met by reducing need rather than trying to supply it by throwing more energy at the problems. It means applying intelligence rather than brute force; start-of-pipe good design rather than an end-of-pipe botched fix.

I happen to take a close interest in energy obviation as it relates to the construction industry. Through a strange set of circumstances that followed on from having brokered the withdrawal in 2004 of a French multinational, Lafarge, from a proposed 'superquarry' scheme on the Isle of Harris, I now serve, for my sins, as an unpaid member of my former corporate enemy's sustainability advisory panel! Full details of this are on my website, including a 2008 paper that I wrote with a Lafarge vice president debating the mutual tensions and benefits of such collaboration between corporations and environmentalists.[13]

Lafarge is the biggest producer of cement in the world, controlling nearly 10% of the global market. Because cement manufacture accounts for at least 5% of world CO_2 emissions, this one corporation, which operates in seventy-five countries of the world, is the critical link in the chain that accounts for half of 1% of the CO_2 component of global warming. Each year it emits double the CO_2 of Switzerland!

What, if anything, then, can be the saving graces of such a company? Working in partnership with WWF International, Lafarge has now taken costly steps to be an industry leader in reducing all levels of environmental impact. In this it has set a standard for the entire industry. Emissions per tonne of cement, as independently audited, have come down by 14% on 1990 levels. Their target is to achieve a 20% reduction by 2010. This is not easy as they continue to acquire filthy plants in places like China and it takes time to bring them closer to European standards. Lafarge's CEO, Bruno Lafont, has given a formal assurance to all of us on his Sustainability Stakeholders' Panel that he will 'identify innovative solutions for developing sustainable, carbon-neutral buildings which are respectful of the environment'. Personally I suspect that complete and authentic 'carbon neutrality' is a pretty tall order, but like the stars by which we navigate our course, it's a good reference point to hold the eye to. As 'critical friends' of the company, those of us from NGOs, ethical finance houses and trades unions who sit on the Panel consistently push Lafont and his colleagues to go further, aiming to 'make more with less'.

Their credibility rests partly on our nod. Our presence and the credibility it brings helps the company's executives to justify the high cost of their programme to sceptical managers and shareholders. In 2007 it helped Lafarge win the accolade of inclusion in the prestigious FTSE4Good Environmental Leaders Europe 40 Index – which, in turn, helps the company to raise ethically screened capital.

Does such participation compromise the integrity of those of us who are involved? Is it complicity in 'greenwash'? For myself, and after close consultation with my former superquarry campaigning colleagues (all of which is published on my website), I took the view that, as a user of construction industry products, I'd rather have a direct hand in helping to reduce Lafarge's share of global CO_2 emissions than to get uptight about the risks of engaging positively with them. This is the uncomfortable truth of the matter: whether we like it or not, if most of us look into the corporate mirror we'll find our own faces reflected back. It's there in the stuff we buy or use, and we drive it, competitively, whenever we seek out the cheapest products in the shops. My post-superquarry constructive engagement with Lafarge is not an easy position to accept. Not least, it involves me in six days of unremunerated work a year, but it does give deep insights that break beyond the 'green bubble' in which it's otherwise so easy to float, self-satisfied but ineffective.

At the end of the day, I don't believe that 'green capitalism' is the ultimate solution. I think that capitalism carries an intrinsic selfish dynamic that militates against right relationship with one another and the Earth. It is a system by which money is the primary guiding hand that shapes decisions and provides targets for motivation. Law and ethics usually do provide a framework of constraints to this, but the ideals that these encode are rarely the prime motivators. This leads to an asymmetry of power. It means that while God always has the nice ideas, the Devil has the bigger moneybag with which to force his wicked way!

My choice, as we will see, is to value mutual business structures more than capitalist ones as the humane and sustainable way to organise economic affairs. But until the building blocks for that are more adequately in place – until I can buy my cement or roofing tiles in the local Co-op – I have to concede that it is at least encouraging to see the social and environmental efforts of an enterprise like Lafarge.

Let me go even further at this stage. Many of my mentions of 'industry' in this book are, for good reasons, neutral or negative. That is because I am building up to a critique of consumerism. But when science, technology and industry helps to satisfy fundamental human needs in a dignified and sustainable manner, then we become witnesses to and beneficiaries of human effort and ingenuity that merits high admiration and gratitude for its contribution to society. At its best, this finds application through industrial ecology. It links human needs to the cycles of nature in ways that follow the flow of raw materials and energy – not from the cradle of a mine to grave of a landfill site, but from cradle to cradle – so that the waste from one cycle of activity becomes raw material for the next. When this happens – when, for example, Lafarge uses waste gypsum from power station scrubbers as the raw material for its plasterboard and so reduces the need to quarry virgin gypsum – it is an ecological delight . . . and it often makes sound commercial sense too.

'New Deal' New Technologies

Some major new technologies that could mitigate climate change are now becoming technically feasible, but only on scales that would require huge government investment. We need to think of these as the energy equivalent of America's 'New Deal' under Roosevelt. One example is carbon capture. It would pump CO_2 from power stations back underground into depleted gas fields or other suitable geological structures. Energy would be extracted from one hole in the ground and the exhaust put back down another. Whilst this would cut greenhouse gas emissions, the economics within free market systems do not add up without government push. So far in Britain this has not been forthcoming. For example, BP was to have developed a pioneering carbon capture scheme at Peterhead in Scotland, but in May 2007 British government vacillation forced them to pull out. The risks and costs are too high to expect industry alone to take the initiative on such a scale. That is why the environmental crisis calls for more governance, not less.

Different countries have different opportunities depending upon their geography. In Britain's case, probably the single most effective raft of renewable energy technologies on a new deal scale involve making use of the coastline – offshore wind, wave and tidal power. For example, the British government's

Sustainable Development Commission considers that Britain could get 10% of its electricity from tidal power alone, the biggest scheme being a 10-mile long barrage across the Severn.[14] This would provide 4.4% of Britain's supply at an estimated investment cost of £15 billion. On the surface of it, £15 billion is a lot of money. However, the Severn Barrage would have an estimated life of at least 120 years and the money shrinks into perspective when set against the £25 billion plus that the Government lent to bail out investors in the failed bank Northern Rock, or the £32 billion that is poured annually into the black hole of defence. The bottom line question is whether shoring up the Earth's life support system is not at least as important as shoring up confidence in the financial system, and whether Britain's security needs are not better served by establishing renewable energy capacity than in trying to secure oil supplies from dodgy propped-up regimes in the Middle East.

Gee Whiz Technologies

One approach generating much interest in America is the principle that if you can't stand the heat, turn down the cooker. NASA is presently exploring ways to reduce the intensity of sunlight coming in to the Earth. One technique would be to erect a giant parasol in space to reflect sunlight away. Another, put forward by Professor Stephen Salter of Edinburgh University, is to spray seawater into the upper atmosphere, providing nuclei that would promote the formation of reflective clouds over the oceans of the world. Weirdest of all is the idea of reversing decades of clean air policies and actually adding up to 1% sulphur to aviation fuel so that droplets of sulphuric acid hover about in the stratosphere. These would mimic the cooling effect lasting for several years whenever a major volcano erupts.

Such gee whiz ideas are known as 'geoengineering'. They pose huge questions as to who has the right to control the Earth's climate. Whose benefit would they be calculated to serve, and what about those whose climates might be messed up in the process? For my money, the best soundbite on geo-engineering comes from Stephen Schneider, professor of climatology at Stanford University. As he described it: 'Of course it's desperation. It's planetary methadone for our planetary heroin addiction.'[15]

Nuclear Fission

In his weekly newspaper column in the *West Highland Free Press*, Donald Macleod – the Free Church theologian who shared a platform with me when campaigning against the proposed Harris superquarry – tells how there was a terrible air crash back in the 1950s. People were horribly burned and much debate ensued about aviation safety. Some pundit came up with the clever suggestion that 'the time had now come to institute an urgent quest for non-inflammable fuels'! Professor Macleod suggests that we are presently witnessing a similar wild goose chase for sustainable fuels.[16] We're not willing to downsize our lifestyles to accord with the output of small-scale renewables. We're being told that we can't keep pouring carbon into the atmosphere with impunity. We don't like big windfarms because they spoil the view. Tidal barrages would spoil the beaches. And nuclear is beneath contempt. So bring on sustainably non-inflammable fuel!

Against the backdrop of that conundrum, let us, then, take a look at nuclear fuel. At the time of writing the most recent British government data, published in January 2008, covered the years 2005 and 2006.[17] In 2005 nuclear provided 20.5% of all UK electricity and 37.9% in Scotland. In 2006 the UK figure drops slightly to 18.9%, but Scotland falls sharply away to just 26.4%. Was that because renewables were rapidly replacing nuclear, as some hopeful observers presumed? Sadly not. It was simply that it had been a bad year for shut-downs with major repairs at both of the Scottish reactors. It also demonstrates how carefully we have to watch out for statistical quirks, especially where they concern issues embroiled in controversy. The bottom line is that nuclear currently supplies nearly 40% of the electricity produced in Scotland and 20% in Britain as a whole. That latter figure represents 8% of Britain's total energy requirement when non-electrical sources such as coal and oil are factored in.

Generation of Electricity by Fuel in Scotland and the UK, 2005 and 2006

	2005 UK	Scotland	2006 UK	Scotland
Coal	33.9%	24.7%	37.7%	32.7%
Oil	1.3%	3.9%	1.3%	4%
Gas	38.3%	19%	35.5%	21.7%
Nuclear	20.5%	37.9%	18.9%	26.4%
Hydro	1.2%	9.3%	1.1%	7.9%
Renewables	3%	3.8%	3.4%	5.1%
Other	1.8%	1.3%	2%	2.3%
Totals	100%	100%	100%	100%

Source: BERR, January 2008,
http://stats.berr.gov.uk/energystats/etdec07.pdf

On the surface of it, nuclear energy is therefore enormously attractive. If just two commercial plants – Torness in the East of Scotland and Hunterston B in the West – can provide such a high proportion of electricity in a country of 5 million people – and a fair chunk of that is exported to England – why are we phaffing around with unsightly windfarms and eco-devastating coal? After all, the French generate 76% of their electricity from 56 nuclear plants and seem to circumvent most of the problems that we perceive. James Lovelock of Gaia fame even argues that we should radically increase the contribution of nuclear as a 'soft landing' transition to a low-carbon future. Before Chernobyl, I would have agreed with him. But Chernobyl reminded us that one definition of an unacceptable risk is an uninsurable risk, and the nuclear industry has no insurers willing to cover major accidents or terrorist attack. The trouble with the once-powerful insurance argument today is that it therefore puts us in a cleft stick. Climate change caused by carbon emissions is just as much an uninsurable risk, so stalemate!

Nuclear fission may therefore be 'the answer'. Reactors designed for 'intrinsic safety' using, for example, 'pebble bed' technology, may be the way to go. Spent fuel rods could be encased in cylinders of steel and copper, cushioned in clay and placed in stable rocks deep underground: scientists at the Finnish nuclear waste disposal authority, Posiva, think that this would make high-level waste safe even through earthquakes and ice ages.[18] But before jumping at any of these and assuming that nuclear is indeed the answer, what was the question? If the question is: 'how can exponential economic growth be sustained on the back of an energy intensive economy?' and if that's what people democratically vote for, then I can see no alternative to nuclear in the short to medium term. Such is the reasoning why, in January 2008, the UK government announced a new generation of nuclear development. But if instead our question is: 'how best can we satisfy basic needs for warmth and other aspects of wellbeing?' then energy obviation becomes a more obvious first step.

Let us look at some figures behind that statement. Apart from the risk of accidents, a problem with nuclear energy is that it panders to the buy-now-pay-later mentality of drip-feed on the credit card. The decommissioning costs alone of Britain's existing nineteen nuclear sites are presently estimated at £73 billion over the next 100 years. From this we can make a revealing calculation. There are 24.7 million households in the UK. That means that our £73 billion – and it's not counting the costs of power station construction, operation and waste storage – works out at £3,000 per household! The Department of Trade and Industry says that insulation standards in the average British home scores a rating of just 45 compared with 100 to 120 at the 'highly efficient' end of the scale.[19] If £3,000 per home were spent on insulation or solar water heating panels, a massive improvement could be achieved. It would free up power station capacity that presently disappears through people's roofs. It would permanently cut domestic fuel bills and especially favour the poor. Is that not a programme that could be given political legs?

Obviously the existing nuclear power plants would still have to be decommissioned. The question now facing us is whether we want more of them. And that's about my bottom line on the nuclear question. The answer all depends on what kind of a society we collectively vote to become. My vote is for a wholesale social shift towards low-impact simple living. Depending on how far it went, 'Nuclear power, no thanks!' could be consistent with that position. But as matters presently stand, the question of how luddite we may or may not want to become is an academic one. Fewer than 5% of the British population vote for radical environmental politics. Most vote mainly by their hip pockets. And while that continues to be the case, nuclear power probably remains, at least in the short to medium term, the best in a devil's pack of cards.

Nuclear Fusion

But let's not yet give up completely on the quest for non-inflammable fuels! Nuclear fission splits atoms in a dirty and intrinsically risky process. In contrast, nuclear *fusion* releases energy by combining them in a physical process that is relatively clean and safe. In principle, fusion holds out the hope of one day providing nuclear energy without the massive waste disposal problems of fission. It would mainly use hydrogen isotopes that occur naturally in sea water. Even if a fusion reactor got hit by a hijacked aircraft, it would just fizzle out rather than melt down. The nuclear critics should repent! Non-inflammable salvation is nigh! There's just one problem. It remains verily on high. While fusion suitable for electricity generation has already been successfully demonstrated in the lab, it is still a long way off becoming commercially viable. At present more energy has to be put in to get the reaction going than can extracted back out. Until that balance changes, commercial fusion remains a hope, but not a reality.

At the time of writing the world is committed to spending just £6.6 billion on the International Thermonuclear Experimental Reactor being built at Cadarache in France. This pools the resources of the EU, the US, Japan, Russia, China, South Korea and India. But that is only £1 for each person on the planet. In comparison, the lifetime cost to Britain alone of the Trident nuclear submarine weapons programme has been estimated at £30 billion. As a nation, we need to get our priorities straight. We should scrap Trident and pump the money into true security – both renewables and fusion research.

I am acutely aware that in strongly supporting the exploration of nuclear fusion I am coming down on the side of a 'technofix'. However, given the way that the world and its population level is presently structured, I do not see how we can turn our backs on this possibility and still take mainstream electorates with us. It is also pleasing that fusion research is being conducted on a basis of international cooperation rather than competition. While nuclear fission disturbs me on a number of levels, I confess that the prospect of fusion generates a thrill of excitement and even optimism for human scientific ingenuity. But that mustn't leave us dropping all our eggs into a basket that doesn't yet exist. Ever since I was a boy, nuclear fusion has been 'thirty years down the road' in the long grass. The grass looks shorter now than it once did, but there's still a good thirty years' worth to scythe a way through.

The Solar Energy Internet

While my dismal review of energy options presented here is far from exhaustive, I would like to close with a brighter ray of hope for renewables, albeit one that would require the development of a massive infrastructure. With a suitable international grid system, the hot deserts of the world could export vast quantities of solar power. It could be gathered either by photovoltaic panels or using mirrors to focus the sun's energy. Already in Andalucia in Spain, one such solar station produces 11 megawatts – enough for 6,000 homes. A field of steel mirrors focuses sunlight onto a boiler positioned up a forty-storey concrete tower. The steam thereby produced drives turbines that generate electricity. The Spanish hope that, as it expands, this plant will eventually supply enough power for the equivalent of the 600,000 people who live in Seville.

With a direct current electrical grid connecting up nations, solar energy could in principle be captured in North Africa and exported to Europe. America could use its own deserts. While the solar internet would require backup generating provision for night and bad weather, the energy could be stored – for example, in pools of molten salt – and released when needed. Because deserts are not agriculturally productive they make a better location for solar collection than the densely populated farmlands of Europe. As such, nations on the edge of the Sahara could provide us with renewable energy. We in turn could provide them with food. The prosperity so created would at least be some compensation for the climate change our affluence visits upon them.

The reason that large-scale solar collection has not yet happened is that, so far, energy prices have not gone high enough to provide the incentive. The good news is that they're getting there, and the potential is huge. Direct sunshine delivers 1.3 kilowatts per square metre. According to the International Energy Agency, photovoltaic panels placed over half of the world's major deserts would produce eighteen times more energy (which is 216 times more electricity) than the world currently uses. In other words, present electricity demand could be met from just one quarter 1% of the deserts.[20] But for developed European nations to contemplate this would require a very different relationship with our neighbours in North Africa. We would have to rethink our links with the Muslim world and explore together, for example, shared ethical and spiritual principles that could work towards the common good. This is not something that either side could enter into without deep self-examination. That is what makes it such an exciting prospect.

A massive and potentially highly practical example of energy from the deserts that is already starting to unfold is the $22 billion Masdar Initiative launched by the government of Abu Dhabi in 2006.[21] Between 2008 and 2016, the government of Sheikh Mohammad bin Zayed Al Nahyan is constructing Masdar City which, it claims, will be 'the world's first zero-carbon, zero-waste, car-free city', growing eventually to support 1,500 businesses and 50,000 residents powered mainly by solar energy. As such, a small island state in the United Arab Emirates hopes to change from being one of the world's largest exporters of oil to one of the largest exporters of solar energy and its associated technologies. It remains to be seen whether Masdar will turn out to be, as its critics put it, 'just a fig leaf for the oil-rich Gulf emirate . . . a luxury development for the rich,' or a groundbreaking demonstration of how the energy basis of a Westernised lifestyle can be turned on its head.[22] Whatever is the case, it is a strange and not unpleasing irony to think that the very nations whose oil from under the deserts has driven global warming could start to provide a solution by no longer digging a hole for us all!

One last point about electricity is that its ready availability obscures the fact that it is very 'high grade' energy. That means it is highly adaptable in ways that allow it to be used with great flexibility. In contrast, primary energy from carbon is relatively low grade. Often we use it to produce heat that does other things, like powering engines or generators. To raise the energy grade of a fuel like coal requires conversion to gas, or to electricity via a power station – all of which is cumbersome, costly and not very efficient. In weighing up the relatively high costs of solar electricity it is therefore worth remembering that it directly delivers energy at a high grade. As my colleague Ulrich Loening points out to his students, all of us regularly demonstrate our willingness to pay very high prices to have tiny amounts of high-grade electricity delivered in the right places: we do so whenever we buy a battery. A typical AA-size battery such as might power an MP3 player provides (at best) in the region of 2,000 milliamp hours at 1.5 volts. It therefore yields three watt hours of power. If we buy, for argument's sake, three of these batteries for a pound, we're therefore buying nine watt hours of energy – call it ten in round numbers. We'd therefore need to spend £100 on such batteries to produce one kilowatt hour of power, which is to say, one 'unit' of electricity. But the cost of a unit of electricity from the mains supply is only about ten pence on our household electricity bills! This demonstrates that to get our high-grade energy in the right places, we're willing to pay 1,000 times more than the domestic tariff.

The same principle can be seen reflected in our household energy bills. My own electricity, which is on a 'green' tariff, presently costs 9.3 pence per kilowatt hour. My gas bill, however, charges only 2.2 pence for the same amount of energy. The electricity is therefore roughly four times more expensive than the gas. Unless my gas central heating system runs at only 25% efficiency (and hopefully it will be well over double that), I'd be a mug to use electricity for anything other than spot heating in the house. I do that, for example, where I have a small (500 watt) electric heater in my home office to avoid putting on the entire central heating system when working alone.

Such is the sort of energy awareness that needs to be made a part of every child's education if we are all to learn to live more sustainably. The general principle is that people need to develop a feel for energy grade, its units of measurement, how much is required for different uses and what makes that use efficient or otherwise. High grade energy should be conserved for high grade purposes, and low grade used for low grade applications like the warming of space and water. As for patio heaters – these should be consigned to the Great Cosmic Downunder and specially reserved for roasting their erstwhile . . . but maybe we'll hold back on the heavyweight theology until Part 2 of this book! Suffice for now to conclude that there are no easy answers to the energy problem. If we want to sustain and even intensify a high energy way of life, we are beset at every turn by devil's dilemmas. Every measure we can think of has its downsides, and some, like growing biofuels – one of the great green hopes not so very long ago – may even yield net environmental losses. Such is the consequence of a world in which so many people strive to live at such a high level of affluence. Short perhaps of a massive technical breakthrough in nuclear fusion, there is no long term alternative but to learn to live within the Earth's carrying capacity.

* * *

The need to link energy security to a wider advancement of convivial international relations that we have touched on in this chapter is why the Scottish Green Party gives its four interconnected principles of sustainability as being: ecology, equality, radical democracy and peace and non-violence. Whatever one might think of a particular political party, those, surely, are the kind of principles that need to be introduced at all levels of politics. The emphasis on peace and non-violence is crucial. In many ways I consider it to be the most important step. We will see why this is in Part 2 where we examine violence as an underlying dynamic that drives consumerism, allied to such drivers of reality distortion as vanity and untruth. But before we broach such metaphysical ground we need, first, to exhaust our consideration of whether conventional drivers of social change have the power to tackle the dilemmas that face us.

Thus far we have seen that the world as a whole needs to cut its greenhouse gas emissions by about two-thirds to stabilise climate. On a back-of-an-envelope basis, one third of this could come from energy conservation and the other third from renewables. The final third can be the carbon economy's business as usual, because the Earth can sustainably absorb that much impact. Achieving that three-thirds scenario is the great green hope. But for this to have any prospect of coming about, humanity's aggregate level of consumption first needs to be stabilised. The trouble is that most of the change needs to come from the rich. That's the difficult bit, because *we* Westerners are 'the rich' in global terms. That's what creates the imperative that Britain, and other nations like it, urgently cuts carbon emissions by about 90%.

Is that a likely prospect? This is the political question to which we must now turn, and I must warn my reader that my assessment is dour.

Chapter Four

SPIRIT OF THE BLITZ

Most people were astonished when John Major was put back into 10 Downing Street after the Conservative Party won its fourth consecutive victory in the 1992 General Election. The polls had confidently predicted victory for Britain's old-style Labour Party under Neil Kinnock. Such was the margin of error that the pollsters' own Market Research Society launched an investigation into what had gone wrong. An average of more than fifty pre-election surveys of public opinion had pointed towards Labour leading by 1.5 percentage points. As events had it – 'events, my dear boy', as Macmillan described the greatest challenge of political life – the Conservatives won by a whopping 7.6 percentage points.[1]

It was an election more than any previously fought in Britain that was won on the principle that 'it's the economy, stupid'. The promise of tax cuts had been Major's secret weapon, not because it was a secret policy but because it reached through to secret places in the psyche of the electorate. The hard truth for Labour was that a significant proportion of swing voters had misled the pollsters. They said they aspired to Labour values, but when it came to the crunch they voted by their hip pockets.

Tolstoy said it all in a much-lauded aphorism: 'I sit on a man's back, choking him and making him carry me, and yet assure myself and others that I am very sorry for him and wish to ease his lot by all possible means – except by getting off his back.' I'm reminded, too, of a cartoon of Robin Hood. There's Robin with his sword drawn, valiantly combating the Sheriff of Nottingham on the castle stairs. But a beautiful young lady shelters behind the sheriff. And the caption reads: 'Maid Marion told Robin that she loved the poor, but had decided to stay with the rich.'

That's the way it also seems to be with climate change. The report of a 2007 Mori poll, *Tipping Point or Turning Point?*, makes disturbing reading. It found that 88% of the British public think that climate change is taking place, but only 41% believe that humans are even partly responsible. Seven out of ten people polled think the government should take the lead in combating climate change, but only 21% support increasing the cost of flying, and a mere 14% would endorse upping the cost of petrol.[2] Mori concludes 'that we are willing to take action and "do our bit", but so long as it doesn't intrude or impinge on the most important aspects of our lifestyles'. We will buy long-life light bulbs and recycle our bottles, but 'the idea of giving up cheap foreign holidays is simply non-negotiable at the current time'. The report merits quoting at some length:

> Past and recent trends certainly give little cause for comfort: annual carbon dioxide emissions are now only 5.3% lower than in 1990 and have actually increased by 2% since 1997; energy consumption in the household sector has risen by about 40%; distances travelled by private car increased by 17% between 1996 and 2004; and the number of passenger kilometres by plane rose from 125 billion to 260 billion worldwide between 1990 and 2000 . . .
>
> Turning to behaviour, the environment has long been a litmus issue when it comes to the clash between individualist consumerism and wider world citizenship, and our research indeed finds conflicting and competing mindsets. The public are keen to protect their own individual lifestyles and choices but, at the same time, appreciative and supportive of the need for change. They look to Government and business to act on their behalf,

> but aren't always so sure when a specific policy or price premium looms into view.
>
> This pattern is evident throughout the climate change debate. The public want to avert climate change and play their part but at the same time they also want to go on holiday, drive to work, own a second (or third or fourth) home and buy the latest electrical products. This climate change equivalent of Orwellian Doublethink, or cognitive polyphasia, does not mean the public don't care about the environmental consequences, but rather, for certain behaviours and en masse, they don't care enough. They hope for technical innovations or efficiency improvements – such as airplanes and cars that don't emit CO_2 – rather than contemplate radical changes in lifestyle.

I, for one, didn't know what 'cognitive polyphasia' meant, and when I searched the web, people were saying they'd first heard of it from Mori! An article in *The Guardian* says it's 'the ability to hold conflicting ideas about the same thing' – and that seems to describe our conundrum perfectly. Governments are at the cutting edge of people's expectations. They catch the collective projections of what we are in denial of and therefore carry the brunt of compartmentalised thinking. Politicians are therefore under huge pressure to play with sleight of hand in environmental reporting. For example, Britain claims to be on target for cutting emissions under the Kyoto Protocols. But as we have seen, this has mainly been made possible by the painless switch from coal to cheap and clean gas.

Since 1990, Britain's total electricity generating output has risen from about 300 to nearly 350 billion kilowatt hours.[3] During 2004 alone – the most recent year for which I could find full data – the UK consumed 1.9 million barrels of oil per day which was a 7.9% increase on the previous year. A report from the UK Office of National Statistics shows that between 1971 and 2001, domestic energy use per household rose by 5%. It went up from 1.87 tonnes of oil equivalent per home to 1.96 tonnes. That is a modest rise. Increased affluence was paid for by increased efficiency. But the rise in the proportion of people living in single households meant that the total domestic energy consumption rose over this period by a staggering 36%.[4]

Meanwhile, our politicians utter reassuring sniffles and snuffles. They do so not entirely because they are stupid, but because, unconsciously, we the electorate set them up for it. We in our cultural narcissism expect them to keep cornucopia flowing while, at the same time, telling us we're good children whose behaviour is slowly getting better. For example, the 2007 climate change report of Scottish ministers details many impressive initiatives that suggest we're rising to the challenge of climate change. Strategies have been drawn up sector by sector for transport, forestry, energy, and so on. And the CO_2 emissions figures look quite impressive – except that, like the figures for Britain as a whole, they omit our share of international aviation emissions. This sleight of hand is justified on the grounds that these take place outwith our own territory! In other words, carbon footprint is conveniently measured when at home but not abroad.

Worse than that, under the EU's emissions trading scheme the UK government is proposing to buy foreign credits to meet fully 70% of its required greenhouse gas reductions. In August 2007 the House of Commons' and House of Lords' Joint Committee on the Draft Climate Change Bill reported on this with rare and exemplary directness. It said, 'The bill as currently drafted would still theoretically allow all the savings to be made externally to the UK, notably in developing countries, and thereby postponing the decarbonisation of the UK economy.'[5] The position is no better in Scotland. In June 2007 the Scottish National Party government said it was starting consultations over a Scottish Climate Change Bill. It would target an impressive 80% cut in emissions by 2050. But these aspirations will not be made legally binding. At the time of writing, it is expected that they will be reported on only once every five years – perfect 'not-in-my-term-of-office' political NIMTOism![6] As such, they fall seriously short of bold SNP manifesto commitments for 'mandatory carbon reduction targets of 3% per annum'. It exemplifies the 'cognitive polyphasia' or as I would prefer to call it, using the closely related term, 'cognitive dissonance' with which we are challenged – pious pledges out one side of the mouth and business as usual from the other.

The 2007 annual report of Scotland's Climate Change Programme hits the nail on the head. It says: 'Addressing climate change will only be successful if everyone accepts responsibility and adopts more sustainable patterns of behaviour.'[7] The finger of government thereby points back at us, the electorate. The trouble is that in a democracy, we only get the politics that reflects who we are.

* * *

For the past four years Vérène and I have lived in the Greater Govan area of Glasgow. Here the shipbuilding industry has ebbed with the tide of Empire and today many of our neighbours live hard-pressed lives. Most people are by no means unaware of climate change and it's mostly not their lifestyles that are the cutting edge of the problem. But in terms of voting priorities, climate hardly registers, and no wonder. As the Mori report says, the public get fed a mixed message. The newspaper of choice in Govan is *The Sun*. When Tony Blair happened to say something sensible at the launch of the Stern Report on climate change, the Scottish edition of *The Sun* of 30 October 2006 ran his picture with a front page headline: 'I'm saving the world . . . YOU lot are paying'. On the inside pages it paraded 'environmental economist' Professor Julian Morris to generate further headlines: for example, 'Families Facing £1,300 Eco Hike [in] Green Tax Blitz'.

Professor Morris is one of those academics who seems to have made something of a career out of pooh-poohing climate change. One wonders how much his interests are really allied with those of people in an area of multiple deprivation like Govan – a place that, for the most part, lies very close to sea level and where the River Clyde is tidal. In 1994, the professor edited a booklet for the right wing Institute for Economic Affairs called 'Global Warming: Apocalypse or Hot Air?' According to the publishers, this claimed that, 'such analysis as has been made by no means supports the view that climate change would place intolerable burdens on future generations.'[8] The professor evidently hadn't changed his mind in the years that followed. He told *The Sun* that because cheap flights, motoring and wasteful domestic appliances would all be targeted by what the Stern Report from the UK Treasury and the Cabinet Office had recommended, '*Sun* readers will be affected most by these changes. Their whole way of life will alter forever.'

Meanwhile, Tony Blair himself was disarmingly frank about climate change. Asked if he would encourage people to give up long-haul holiday flights, he told the BBC: 'You know, I'm still waiting for the first politician who's actually running for office who's going to come out and say it [that people should not fly] – and they're not. It's like telling people you shouldn't drive anywhere.'[9]

Personally I keep a careful eye on indicators of public attitude. When gathered in the company of fellow environmentalists it is easy to feel optimistic. The talk goes like this: we could do this, we could do that, and we'll all be much the happier for so doing! But step outside the green bubble and dig a little into the secret recesses that actually drive average voter behaviour, and a picture often emerges that is not just disappointing, but disturbing.

The 1992 election is one example in this respect. Another, with a specifically environmental focus, was in July 2007, when Boeing launched a new aircraft, the 787 Dreamliner. This was designed to emit 20% less CO_2 than its competitors. The BBC ran a forum asking what people made of it. Would the Dreamliner help to tackle climate change, or would it just generate cheaper flights and more flying? After all, there is evidence of a worrying 'rebound effect': the money people save on energy efficiency gets displaced into other consumption, thereby leaving no net environmental benefit! I read through a sample of the responses that the BBC's Dreamliner question drew from the public. There were many, and more than half made angry attacks on what they variously called 'eco-veggie-commies' and 'enviroNazis'. These luddites were trying to 'push humanity back into the Stone Age' by having the audacity to problematise flying. Some respondents pushed the

Martin Durkin line that climate change is caused naturally by the sun. In any case, who cares if Greenland melts? According to the self-styled 'Mr Average of Southport' (who must have been a *Sun* reader), picking on flying has 'become a class war thing'. Mr Aretheyserious scoffed that 'anything other than living in a cave, riding a bike and planting trees all day will be at odds with environmentalists'. And so it went on in page after page of reaction from those who chose the virtual reality of their computers rather than the actual reality of planting trees.[10]

Similar wails of outrage greeted the Conservative Party's surprisingly commendable Quality of Life Group policy review in September 2007. Of the top twenty reader-rated responses that I looked at on the BBC website, only one of them welcomed the report – and he was ranked number twenty! A typical respondent was 'The Grump' from Chelmsford. He or she complained: 'When will politicians understand that green taxes are still taxes and will not be vote winners? Sorry, I want my taxes down – end of story!'[11] Such eco-loutism is evident on any number of other web forums where people can write in anonymously and thereby dodge the dissonance between what they really think and what it's respectable to say in public. Those who shout loudest, of course, are not necessarily the most representative, and yet they have political muscle. Their coercive power became plain in September 2000, when Gordon Brown's efforts to push up fuel excise duty – in effect, a carbon tax – hit the political buffers of national crisis. The so-called fuel protestors blocked roads and brought the nation quickly to its knees. Brown was forced to concede to their demands, and one suspects from his environmental meekness, that he's still in recovery from the trauma. As one of those car bumper stickers puts it, 'You toucha ma car, I smasha your face!'

As if denial wasn't bad enough, some people positively embrace the prospects of climate change. In September 2007 a BBC World Service poll suggested that 65% of Russians knew little or nothing about global warming, compared with just 10% of British and Americans. Rinat Gizatullin, a spokesman for Russia's Natural Resources Ministry, responded: 'If anything, we'll be even better off: as the climate warms, more of Russia's territory will be freed up for agriculture and industry.'[12] In Scotland, the government's climate change scoping study acknowledges that 'positive impacts from warmer temperatures may include benefits to the forestry and agriculture industries; a reduction in cold-related deaths; and the feel-good factor associated with warmer climates'.

Just how good it could feel was underscored by former geology professor Richard Selley of Imperial College. Now retired, Selley is an authority on the history of wine. Using the IPCC's forecasts, he reckons that in eighty years' time, France and even England will have become too hot for vintage production. Instead, Scotland will be the up and coming epicentre of viniculture. The south-facing north-west slope of Loch Ness benefiting from extra sunlight reflected off the water would be absolutely ideal![13] The celebrity chef Nick Nairn chipped in to the debate with his particular brand of expertise. Speaking at the launch of the Scottish Government's 'national food discussion' in early 2008 he said, 'Here in Scotland we are spoiled for choice . . . in the amazing fruit and vegetables we are growing thanks to climate change.'[14] Meanwhile, Iain MacWhirter – an intelligent commentator but with his tongue here monstrously in cheek – surmised in the *Sunday Herald*: '. . . and house prices in Edinburgh could keep rising – forever. It's all beginning to look rather attractive: a warmer, richer and more populous Scotland, which will be able to regard with pity the plight of millions in the hot world.'[15]

And there's the political rub. The political will isn't there because the electorate in most northern nations aren't worried enough. Across Britain as a whole, we therefore take steps to double airport capacity and anticipate a 40% increase in road traffic by 2025 requiring 2,500 miles of new roads.[16] Figures from the Department of Transport show that the cost of rail travel has risen by 5%, inflation adjusted, since 1997. Bus fares on the same basis have gone up by 15%. But the cost of driving *fell* over the same period by nearly 10%.[17]

Oh dear! It is all so very depressing. I struggle in a losing battle to remain a green hopeful – to retain optimism that the world can dig its way out of this hole. How much does our society really care? How much are we just plain selfish and masters of denial? Ever since I can remember, Third World aid agencies have been passing round petitions urging the government to honour its United Nations pledge to spend 0.7% of GNP on overseas aid. But have we ever done so? Not likely! Britain gives precisely half of what it promised. I can't help thinking that if we haven't been able to bite the bullet with poverty up until now, then what are the chances that we'll take radical action to save the poor as climate change kicks in? We love the drama of a bit of military intervention in the world's hot spots of conflict. But sustained systemic change to constrain our own way of life? That's not the legacy of what put the 'Great' into Britain!

Amongst environmental organisations there is a kind of unwritten pact that we all have to stay optimistic. We have to play the game, at least in public, of pretending that tinkering with a new emissions target here, and changing lightbulbs there, is the way to start tackling the problem. Well, we should be grateful for small mercies, but personally, I've started to find that far from pepping things up, this sort of denial drains my energy. It panders to appearances, but buys into the lie. Green hopefuls will say, 'Ah yes, but remember how we abolished slavery. Saving the environment is of the same order of social change. We did it then and we'll do it again!' Well, let's examine that argument. Slavery didn't create many unemployed slaves who, overnight, signed on for their welfare benefits. Rather, it yielded to waged labour, the main difference being that former slaves now had the freedom to choose their master. Even Adam Smith in his *Wealth of Nations* had said that a surly slave workforce is not the best way to run an economy – except, he conceded, on labour intensive plantations. In any case, slavery was abolished at the same time as a massive transition was taking place towards industrialisation. Yes, it cost the nation a lot, but the economy was on a trajectory where technology would soon make good the losses. The abolitionists were well aware of this – which was why many of their arguments were pitched on economic as well as moral grounds.

One horsepower, after all, is a measure of work equivalent to three-quarters of a kilowatt hour. It would therefore take four horses to power a generator (at 100% efficiency) to boil even just a standard 3-kilowatt kitchen kettle. In contrast, one 'manpower' – an eighteenth-century measurement of the sustained output from a healthy male worker – is just one-twelfth of a horsepower. At just over 60 watts that's merely sufficient to illuminate the average incandescent lightbulb! So much for the output of a slave if measured only in terms of the brute force that interested plantation owners. It would take forty-eight working a treadmill at 100% efficiency to boil a kettle! While the achievements of the abolitionists were staggering and they set the template for much future activism, it's no wonder that technology was nonetheless destined to outpace slave and horsepower alike.

'Then what about the spirit of the Blitz?' ask the green hopefuls. 'Climate change is an even bigger threat to security. If our parents and grandparents accepted austerity under the Blitz, then surely . . .' Well, is climate change really going to hit Britain like the Blitz did? It probably will, long term, as a slow holocaust, and on a world scale. But will it do so within the political time horizons or even in the lifetimes of most of those of us who are around now? Let's remind ourselves of what happened in the Blitz.

The first four days of September 1939 are often described as having seen 'the biggest and most concentrated mass movement of people in Britain's history'.[18] In response to the terse order, 'Evacuate forthwith', issued from Whitehall at 11.07 a.m. on Thursday, 31 August 1939, nearly 3 million people, mostly children and their teachers, packed their bags and fled to places of safety from the bombing. For those who stayed behind, the Clydebank Blitz over just two nights in 1941 saw Luftwaffe bombs kill 528 people and completely destroy 4,000 homes. Between September 1940 and May 1941 Britain as a whole, with London as the main target, saw 43,000 civilians left dead and over a million homes destroyed. The consequences linger on quietly to this day. Only in 2006 did Britain finish paying off its war debts to America. And every night, and perhaps still for another quarter century to come, old folks who never married go to their beds with a prayer on their lips for the loved ones lost in those 'darkest hours'.

'Lest we forget': that's the visceral challenge to the cultural immune system that drove the spirit of the Blitz. That's how pressed to the limits people were. Even then it was a struggle to get everybody to comply with environmental measures to eke out rationed resources. I have in my files a couple of cuttings from the 'Down All the Years' reminiscence column in the *Stornoway Gazette*. One of them, originally published on 7 August 1942, had the caption 'Keep Your Waste Paper Separate'. It reported on the efforts of the Waste Paper Recovery Association – a public agency that we could probably do with reinstating today – and said: 'Housewives and office workers are still not keeping their waste paper separate from other salvage, and this is causing serious stoppages in paper mills engaged in work of national importance . . . Even old boots and bits of machinery were found in the sacks of waste. As a result production in this mill has dropped 15% and damage to machinery has been very serious.'

The other, originally published on 7 March 1947, described how Stornoway's town council was considering a system of 'rhythmic control' for its street lights. This would use an electronic signal to turn the lighting on and off in accordance with need and thereby save energy. The article bemusedly concluded: 'With such a handy system in use, it will be easy for the Council to revert to the old idea of switching off the street lights when there's a moon – an idea which once gained for Stornoway the world-wide fame – or notoriety – of a mention in "Believe it or Not".'

My point is not to disparage such measures as recycling and energy conservation. They are imperative. It is simply to say that even when people are as hard-pressed as they were in wartime, disciplined compliance still takes effort to achieve. The rhythmic lighting story is particularly interesting. It shows that social pressure was making a mockery of frugality even in the early twentieth century. In Chapter 7 we will examine how pressures of that nature were deliberately cultivated to drive consumerism. For now, let's see where it takes us if we face up to the fact that climate change highlights a profound systemic problem. Comparison with slavery or the Blitz only masks the complexity of that problem in simplistic rhetoric.

* * *

In the summer of 2007 public awareness in Britain received an abrupt reveille. England was gripped by the worst floods ever recorded. The first half of the summer was the wettest in the more than 240 years since records had been kept. Paradoxically, as accords with the unpredictable turbulence predicted by climate change models, some parts of Britain – most notably the west of Scotland – were uncharacteristically dry. But down on the southern floodplains thousands of people were washed out of their homes and tens of thousands lost electricity supplies. In Gloucestershire, a third of a million had no mains water for up to a fortnight, ironically because the water treatment plants had been inundated. The RAF mounted what it described as one of its biggest ever peace-time rescue operations. In some places 5 inches of rain fell in less than a day. Even the best flood defences were overwhelmed. Parts of Sheffield, Doncaster, Hull, Oxford, London, Gloucester, Cheltenham, and many smaller towns and villages were left soaked half way up the living room walls.

The cause was twofold. There was, of course, the extraordinary rainfall most probably linked to global warming. But in addition to that we were witnessing the consequences of a general loss of connection between people and place. For decades, agricultural land had been drained, rivers had been straightened, and concrete poured over ground that had once served like a sponge. When the deluge came, there just wasn't enough environmental resilience to soak up and mitigate the impact. Water streamed off surrounding hills and cascaded down distant valleys. Six feet of water surged through the streets of Evesham as if nature was playing a joke: it all took place under a cloudless sunny sky! The actual rain had dropped elsewhere.[19]

Because I am interested in consciousness change I closely

watch the pulse of the nation at such times. The mass media expressed fascination, resignation, anger, and, one suspects, a fair measure of *Schadenfreude* – pleasure in the misfortune of others. It sold a lot of newspapers! From the government, there was talk of more money for flood defences, the need for insurance underwriting, and hardship funds. Lots was said about sticking fingers in dykes and public inquiries into failures in the planning and emergency response system. But there were very few calls seriously to question our way of life.

The BBC's economics editor, Evan Davis, put it all in perspective on his blog. The summer floods, he calculated, would cost the nation something like £5 billion. This 'would still represent under half a per cent of our national income, which is quite small'.[20] During the economic quarter immediately prior to the flooding, Britain had notched up a further 0.8% in its rate of growth. It was a continuation of the uninterrupted exponential growth of the previous fifteen years. To set this against the flooding costs is salutary. The deluge swamps just six weeks' worth, not of economic turnover, but of the mere growth in economic turnover! All the misery that filled the newspapers throughout the summer silly season had hardly chipped the icing on the economic cake. As Hamish McRae estimated in *The Independent*, the monetary costs were nothing more than the difference between an annual growth rate of 2.75% and 3%.[21]

In terms of how GNP measures things, the floods might not even be a bad thing. Each of those insurance claims generates economic activity! We may be more miserable, but on paper we're richer! Never mind the human cost. Never mind, as one flood victim put it, that having to have contractors come in will never make up for the years of loving care he had crafted into his home. In a system of 'positive economics' that only counts what can be counted, progress measures only what's kickable. And just imagine! The uninsured will go into debt, which means they'll have to run even faster on the economic treadmill to keep up payments! That's why debt's good for the economy. It stops the poor amongst us from getting indolent. What doesn't kill the great casino economy makes it stronger.

But too bad for the children neglected by their knackered parents. Too bad for the Chinese who beaver away to export us more stuff. Too bad for nature.

Our economic system is so . . . infantile . . . and it's like that because it's psychologically how most of us are. It's hard to grow up because most of us, most of the time, are sleepwalkers. Charles Tart, the eminent psychologist of consciousness, says that we generally accept the status quo and our powerlessness within it because we're drifting around in a quasi-hypnotic state. He calls it 'consensus trance reality'.[22] It simplifies our cognitive and perceptual processes to help us cope with life. Like someone hypnotised not to see the elephant in the living room, we close down the wider horizons. We accept as 'normal' what our peers consider normal. This is what makes us 'normal'. It's why Tart describes the trance as 'consensual' – something that we all buy into. Social conditioning in the family, education for regimentation at school, suit and tie in the workplace and the psychological onslaught of mass communications like TV and advertising all sustain the mantras that help to keep us 'normal'. I boil myself another cup of tea without stopping to think that a glass of water would do just as well. That's because I too am conditioned. It happens in ways that I see through and, more to the point, it happens mostly in ways to which I'm blind.

The consensus trance helps to maintain a sort of mental health. It helps to keep the bogey man away. We mostly stay 'sane' because the hubris of the world is socially managed within limits we can cope with. The people who can't cope take tranquilisers to reduce their sensitivity to reality. Some drink or smoke or bury themselves in work or in trivia. All these things keep you 'level headed' unlike the wide-eyed prophets, who are always mad. 'Go, go, go, said the bird' in T. S. Eliot's *Four Quartets*. 'Human kind cannot bear very much reality.'

While England's floods were at their worst I tried a mental experiment. The question in my mind was what could politics look like if Gordon Brown (who had just been internally elected by the Labour Party as Prime Minister without having to go to the country) gave climate change the attention it required. At the time Brown's ministers were variously describing the unfolding extreme weather events as 'unprecedented' and 'cataclysmic'. Quite properly, nobody was saying that the floods were sure proof of climate change. At the same time, nobody was ruling it out and everybody had their suspicions. I therefore had the following letter published on 23 July 2007 in *The Herald* – which these days is Scotland's newspaper of record. It was written with tongue only slightly in cheek.

> Sir: As the nation is deluged the government has much to say about flood protection but very little about its underlying causes. I would like to volunteer as Gordon Brown's speech-writer. Confidentially, within the columns of your newspaper, I propose to him the following emergency address to the nation.
>
> 'The evidence suggests that climate change is now the most pressing problem of our times. England's floods are but a symptom of the turbulent future we face. The root causes are greenhouse gases produced by our appetite for carbon-based energy. Action is called for on a scale unprecedented outside wartime. Therefore, I wish to reintroduce and escalate those carbon taxes that the fuel protesters thwarted in 2000. Climate change demands a greater patriotism than that of economic self-interest.
>
> 'With due protection for the poor, we must tax carbon-based energy profligacy until Britain's share of greenhouse gas emissions is consistent with the best scientific advice. The proceeds from these taxes will, first, provide relief for uninsured flood victims. Secondly, they will institute a massive programme of public works for flood protection. And, thirdly, they will be used internationally to mitigate climate change and to compensate those who suffer most: the poor.
>
> 'With other European heads of state, I will require the World Trade Organisation to introduce discriminatory tariffs on trade with nations that would otherwise seek competitive advantage by shirking their responsibilities. And starting with the elimination of nuclear weapons and the recall of our troops from abroad, we will shift resources from the war on terror towards

true security – environmental security – within a new framework of life-giving international relations.

'These are grave measures that must be put to the country. Therefore, I request Her Majesty to dissolve parliament and call a general election.' Yours etc . . .

And did Mr Brown call me? Well, I waited in all day, but he must have tried when my wife was on the phone. Emails of support came in from greens, but the message that I found most instructive was from a person who keeps their finger very much on the pulse of Middle England. She wrote, 'You can't be serious! It beggars belief!'

And that's the conundrum with which I have increasingly wrestled as I researched this book. There is a vast chasm between what needs to happen to avert the IPCC's 'very likely' catastrophic climate change, and what can be achieved from the political bridge of our cultural supertanker. As my English e-mail correspondent went on to say, individual action is like spitting in the wind, 'and I'm not going to spit when I can see which way the wind comes from'.

Meanwhile, over in the South Pacific, the 12,000 strong nation of Tuvalu lies just 4 metres above sea level – at its highest point. One third of a centimetre rise in sea level every year doesn't make much difference to most of us, but to a nation that's as flat as that it spells doom. In 2003 Tuvalu's then prime minister, Saufatu Sopoanga, put it like this to the United Nations General Assembly in New York:

> We live in constant fear of the adverse impacts of climate change. For a coral atoll nation, sea level rise and more severe weather events loom as a growing threat to our entire population. The threat is real and serious, and is of no difference to a slow and insidious form of terrorism against us.[23]

As the climate change campaigner Aubrey Meyer puts it, we need less of a War on Terror and more of a War on Error. No amount of protection from Trident nuclear submarines or even insurance policies can save us from flooding. The problem is that climate change is arriving on a vastness of scale and yet over such a creeping time span in human political terms that we just aren't equipped to respond. In times of war, at least, there is an enemy that can be identified and, hopefully, some sort of an identifiable end point when one side or the other sues for peace. But climate change is a different beast. Inaction will see out most of the present generation of politicians, so why spit in the wind? Why wake the sleeping from their consensus trance? Rome slowly catches fire, but the beers still come across, the girls are dancing, and Nero's got a few more tunes left in his fiddle.

* * *

At the height of the London Blitz, Herbert Mason went out on the roof of the *Daily Mail* building and took his famous picture of St Paul's Cathedral.

The dome rises into light from out of black smoke that billows from the ordinary homes of ordinary people.

And there is a story . . . a story of long ago, that once upon a time two masons worked on a building site. A passing pilgrim asked the first what he was doing.

'I'm just chipping away at another block of rock until payday on Friday,' the man replied.

The pilgrim turned to the second, who said, 'I'm building a great cathedral that will not be finished until long after I'm gone.'

And maybe that's how it is for us today.

Chipping away in a practical way is one part of how we must tackle climate change. That's all the emissions targets and incentives, the taxation and the legislative regimes, the recycling and the energy efficiency, and the technical options. That takes us through to Friday. It fields a short ball on time's playing field.

But without something more, the spirit will run dry. That's why the other part of what we need to do to tackle climate change is to connect with a vastly bigger picture. It means stepping out of the consensus trance of daily grind and limited vision and playing a long ball.

Like that second stone mason – indeed, like Herbert

Mason's iconic picture of St Paul's itself – we must see the cathedral rise from out of the dust of shattered dreams. With eye set resolutely to the breaking of the light, we must become the keepers of cathedrals of the mind.

That is the task to which we must now turn.

Part 2 – The Human Condition

Chapter Five

PRIDE AND ECOCIDE

As we have seen, the Intergovernmental Panel on Climate Change of the United Nations considers it 'very likely' that global warming is primarily anthropogenic – caused by our lifestyles. 'Very likely' means 90–99% confident. We have also seen that the science is complex and, often, contested. It can therefore be difficult for even an informed layperson to figure out what to think and where to stand. That is why, at a risk of 1–9% of looking silly in my retirement, I will proceed from this point on by taking the IPCC's findings as a given. From here on we will leave behind the debates about whether climate change is happening, whether it is anthropogenic and whether it is serious. According to the IPCC, the answer to these is yes, yes, yes. In Part 2 of this book we will turn our attention to even deeper questions. We will explore the implications of the consideration that as well as being a technical, economic, and political problem, climate change is also cultural, psychological and spiritual. It therefore presses us to appraise nothing less than the human condition. We have looked at the facts and figures of the planet's climate in Part 1. Now we must explore the underlying values that shaped how we got there. Most important of all, if our optimism for the future is shaken, we must seek out whether and wherein lies hope.

On the surface of it all, climate change is a completely new challenge. It comes on top of many others including those posed by resource depletion, biodiversity loss, habitat degradation, poverty and war. Most people probably think that worries about climate have appeared fresh in human consciousness. They probably see it as a recent 'emergent property' of industrialisation – that is to say, as a property that only became manifest as the scale of human impact on the planet accelerated. But such a view is far from the truth. Human beings have long worried about being visited in various permutations by hell or high water. Early writers often linked this to messing about with the minerals that nature had secreted away in the Earth's crust. Mining and the smelting of metals has always been associated with air and water pollution, deforestation and avarice. As the Industrial Age took off, people additionally worried about burning the soft black rock known as coal, and now, of course, we can include the effects of mineral oil and gas.

The German scientist Georg Agricola was known as the 'father of mineralogy' on account of his magisterial text, *De Re Metallica* (On the Nature of Minerals). It was published in 1556 and, for nearly three centuries, stood as the West's standard textbook of geology. Its most celebrated English translation from the Latin was produced in 1912 after five years' intense work by a young mining engineer called Herbert Clark Hoover and his wife, Lou, who was both a Latinist and geologist. Later they were to occupy the White House, the Hoover Dam on the Colorado River serving as enduring testimony to Herbert's presidency.

Agricola's copious illustrations plainly show the deforestation and water pollution caused by mining in his day, yet he attacks the criticisms that Roman writers like Ovid, Seneca and Pliny had levelled against such arts. As he saw it, the benefits of extracting wealth from the bowels of the Earth manifestly outweighed the costs. The poets and philosophers were just being impractical dreamers the way that such likes always have been. As a Gaelic proverb puts it, 'You won't find chaff in the poet's byre', meaning that this isn't the sort of character who labours for his bread like everybody else! Agricola found the Roman nay-sayers of mining irritating, but at least he gives fair representation of what they said. Here's how it's put in the Hoover translation:

> . . . they make use of this argument: 'The earth does not conceal and remove from our eyes those things which are useful and necessary to mankind, but on the contrary . . . she yields in large abundance from her bounty and brings into the light of day the herbs, vegetables, grains and fruits, and the trees. The minerals on the other hand she buries far beneath in the depth of the ground; therefore they should not be sought. But they are dug out by wicked men who, as poets say, are the products of the Iron Age.'[1]

The argument between the eco-idyll and technocracy, then, is an old one. But the idea that human wickedness is associated with damage to the Earth is much older than even the Romans. We can find it forcefully expressed in some of the earliest written sources. Common amongst many ancient authorities is the view that 'hell and high water' were punishments sent by the gods who were angered by human hubris – by the 'pride' of ungrounded and inflated ego, wantonness, lies and violence. Discord in the social environment found its nemesis in the destruction of the natural environment and thus in the withdrawal of 'Heaven' from human affairs and the leaving behind of Hell on Earth.

As we shall see in the rest of this chapter, there is a rich irony here. The ancients blamed natural disasters on moral degeneration of which the temptations of mining were but one of many variants. That was certainly a credible diagnosis where there had been local or regional deforestation and loss of soil quality. But the biggest fear of the ancients was of flooding, and that on a global scale that threatened cities near the coast. From a modern scientific perspective this fear probably had little to do with human badness, but much to do with both the plate tectonics of the Earth's crust and, more interesting from our perspective as we will see, to 'natural' prehistoric climate change. The irony, then, is that the ancients developed an astute moral analysis of anthropogenic climate change but one that is perhaps more applicable to us today than it often was to them. As ecological prophets, they were two or three thousand years ahead of their time.

We can therefore be enriched by reading their insights with a fresh eye. However, to do so we first need to develop that eye. We will need, sometimes, to set aside our 'positivist' concern for supposed factual history and understand that ancient manuscripts often deal with the interface between fact and myth. We need to read them less as history than as 'psychohistory' – as a revelation of the interplay between psychology and history. This may be a novel approach because our culture values factual truth over poetic truth. When it comes to seeking knowledge, we are very much in the head and very little in the heart. It makes us strong on empirical facts but weaker in wisdom. In contrast, the ancient mind, like that of many indigenous peoples still hanging on today, understood the value of mythopoesis. They could see that it is not just objective fact, but also forms of truth only communicable in subjective story, poetry and song that matter. These are what shape our experience of and participation in the ongoing birth of the world into reality. Such *poesis* is the root of our word, poetry. It means 'the making'. Mythopoesis is therefore the upwelling of reality from deep springs in the psyche of the world. When we understand Big Bang and evolution in terms of 'the Creation' as comprising such ongoing process, we bridge science and theology by grasping, late and last, the true nature of poetry.

The totality of human experience is 'outer' life harmonised with the activated 'inner' life that it takes to see this. What matters to achieve inner awakening about the significance of human affairs is less historical accuracy than 'psychological realism'. It's about permitting 'hi-story' in the magical as well as the literal sense – in terms of what the literary world calls 'magical realism'. Meaningful truth therefore stands out from morose data collection. It must be not just logically valid; it must equally 'feel' true. Black Elk, the Oglala Lakota (Sioux) warrior and holy man put it perfectly. He said of his people's traditions, 'This they tell, and whether it happened so or not I do not know; but if you think about it, *you can see that it is true*.'[2]

When fact and myth are integrated, the 'head' and 'heart' can open up realms of human knowing we never could otherwise have imagined. Such is the epistemology – the theory of the structure of knowledge – that can expand human experience into new paradigms of insight, and such is the basis on which I now venture to proceed in approaching early historical sources.

Let us start with an ancient text about the climate going crazy that will be familiar to many readers. But this time, let us consider it at least partly as psychohistory rather than incredulous 'fact'. I'm referring here to the sixth chapter of the Book of Genesis in the Bible. Here we have a piece of ancient literature that scholars consider to be about 2,500 years old. I'll present it using the King James 'Authorised' Version of 1611 because that's the most poetic, the most dramatic and, therefore, the most psychodynamic English translation:

> And God saw that the wickedness of man was great in the earth, and that every imagination of the thoughts of his heart was only evil continually. And it repented the Lord that he had made man on the earth, and it grieved him at his heart. And the Lord said, I will destroy man whom I have created from the face of the earth. . . . The earth also was corrupt before God, and the earth was *filled with violence* . . . And God said unto Noah, The end of all flesh is come before me; for the earth is *filled with violence* through them; and, behold, I will destroy them with the earth. Make thee an ark . . . and, behold, I, even I, do bring a flood of waters upon the earth, to destroy all flesh . . . and every thing that is in the earth shall die.

The rest, as they say, is history. But we miss the point if we think this is a tale literally about the animals going in two by two. The point, at least as it serves our purposes and as I've emphasised in italics above, is that we're twice told that it is violence that's the problem. Whatever else the story may or may not be, it most certainly serves as a moral fable. It reveals the worldview of a society that allows violence to inundate its psyche. It shows how that psyche metaphorically floods, and so undermines the entire basis of its own subsistence.

The ancients were able to draw this lesson from what were probably residual memories of actual natural disasters. The connections they imputed to cause and effect may have been debatable depending on the kind of natural disaster that was in question. But if the ancients over-pathologised their psychic condition, we today go to the opposite extreme. Too many of us refuse to acknowledge and act upon the mindlessness that underlies anthropogenic climate change. We forget that hubris, which is a state that we accept almost as a norm in our helter-skelter hustle and bustle society, is a word that has its origin in the Greek *hybris,* meaning 'wanton violence'. We forget, too, that 'God' should be understood not as some fulminating Father Christmas figure in the sky. Rather, God as the theologian Paul Tillich put it is the 'ultimate concern' for which we yearn, the 'ground of being'. As such, we conveniently miss the point if we think that the finger or voice of God stands somehow apart from who we really are. The denial of God so understood becomes denial of one's deepest Self. It is nihilism's last call.

* * *

Noah, of course, emerges from the ark with his family and prime breeding stock with which to replenish the world. Humanity is mandated to fill (but it never said to overfill) the Earth. Meanwhile, God has set the rainbow in the sky as a sign of hope and promise that the Earth will never again be knocked back to Year Zero. Maybe not, but that doesn't mean that, on a lesser scale, wayward humankind might not still have it coming! One man who feared just that was Noah's great-grandson, King Nimrod.

Nimrod was the world's archetypal egomaniac – an exemplar of narcissistic male power. Just so that we don't miss the point, the writers of Genesis gave him a name that means 'rebel'. The point to grasp about narcissism is that the narcissist is not in love with his true self. He doesn't know his true self because he's never done inner work. He only knows a reflection or image of himself that ego identity has outwardly crafted. Inauthenticity lies – in both senses of that word – at his very core, because we're dealing here with a self-centred rather than a centred self. Look in any male fashion magazine to see the type. Here is the man resourced not from the wellsprings of life within, but from the attention – favourable or unfavourable; it makes little difference – that others give him.

Psychotherapists call this attention seeking the quest for 'narcissistic supply'. Adult narcissists – men and women alike – are 'energy vampires' who supply their emotional deficiencies by drawing on others. They're the office egomaniacs, the actual or wannabe bigshots, those bustling centres of constant fluster where everything's on the outside – everything's so jolly, ever-so – and you wonder how much is really there inside. In the primary narcissistic phase of childhood it's perfectly normal and appropriate for the two-year-old to play at omnipotence; perfectly normal to shout, 'Look at me, I'm Superman!' But when this stage of development 'hangs' and carries over as secondary narcissism that persists into adulthood, it becomes socially problematic and spiritually perilous. It may be valued as a competitive stimulus in many of our commercial enterprises, sporting institutions and political organisations. The Donald Trumps of this world may think they've earned our adulation for their self-styled Trump Towers and their trumped-up business plans that succeed on the basis of push, push, push. But really, theologians have a name for such misplaced expectation of veneration, and our celebrity-obsessed culture maybe needs to wise-up about getting sucked into its gameplay.

The Book of Genesis makes the psychopathology very plain.[3] We're told that Nimrod built his tower 'with its top in the heavens' precisely because, as he and his fellow Babylonians are quoted as saying, they wanted to 'make a name for ourselves'. Here, then, we have people who are stuck in the first half of life – in carving out an ego identity in the 'outer' world. They haven't moved on to the second half of life – grounding that identity in the far-reaching relationality of community, and deepening to a fullness of 'inner' or spiritual realisation. They know themselves as the cork that visibly bobs along on the surface of things but haven't yet realised that they're also integral to the river's deep flow.

As Nimrod saw it, his relationship with God was just another battle between egos. His aim was therefore to trump the Lord. In terms of Oedipal psychology, he wanted to give Dad a drubbing, for to wrestle with 'the Lord' at this level is usually a parental projection. That's what gives God such a bad name! 'God' becomes Mum and Dad, and they were never good enough. Conversely, Mum and Dad catch the child's unconsciously protected yearnings for God, and so they're never good enough either! All of that is fairly normal, but the nascissist, lacking grounded relationship to deep reality, acts like they are God. Their 'ultimate concern' has shrunken to themselves. Such are the dynamics when the relationship between the small self and the great Self is out of kilter; when the cork of the ego floats not, as Shakespeare puts it in *The Comedy of Errors*, 'obedient to the streame', but in arrogant ignorance of the wider forces that actually sustain life.

So it was that Nimrod ventured out, Genesis tells us, to make his name. He and his cronies served self rather than others. They did so in the time-honoured manner of building a magnificent carbuncle of architectural phallocracy . . . that 'award winning' apogee of design both urban and urbane, the original Tower of Babel™.

Now, Nimrod venerated the cult of violence. In fact, we're told that he set the performance indicators on this score: he was 'the first on earth to become a mighty warrior'. He began his kingdom at Babel, which was known as Babylon to the Greeks. The name means 'Gate of God'. But depending on what is in their hearts, they who seek proximity to the gods are rewarded with either music or madness. As such, it's revealing that there's a double meaning in the name. Babel is also cognate with 'babble', meaning 'confusion'; and so the word carries a general suggestion of hubris. This, in the eyes of the ancients, was the greatest of sins.

Because the concept of hubris is so central to the thesis of this book, it is worth further unpacking its meaning with some practical examples of usage. A fine one comes from *The Sunday Times* of 16 March 2008. Here Lord David Owen, formerly both a physician and a British foreign secretary, applies the term to George Bush and Tony Blair in the run-up to the 2003 invasion of Iraq. He explains:

> In ancient Greek drama, a hubristic career proceeds something like this: the hero wins glory and acclamation by achieving unwonted success against the odds. The experience then goes to his head: he begins to think himself capable of anything. This leads him into misinterpreting the reality around him and into making mistakes. Eventually he gets his comeuppance and meets his nemesis, which destroys him.

The Oxford English Dictionary similarly cites Aldous Huxley as saying: 'The Greeks . . . knew very well that hubris against the essentially divine order of Nature would be followed by its appropriate nemesis.' It also lists the first known English language application of the word as having been surprisingly recent. It was a newspaper article from 1884 which said: 'Boys of good family, who have always been toadied, and never been checked, who are full of health and high spirits, develop what Academic slang knows as hubris, a kind of high-flown insolence.' That sounds like Nimrod! And as we'll see shortly, it most certainly fits his near-contemporary, Gilgamesh. In short, hubris is the pursuit of life out of relationship with God. The 'obedience' to the stream in question is not blind compliance, but rather a readiness to go with the deep flow of reality – the movement of the Tao, the unfolding of Dharma, the cosmic play that Hindus call *lila* and which courts as lover the Lord or Goddess of the Dance.

That is why theologians consider all else to be idolatrous: 'There is no god but God.'

According to archaeologists, Babel was located 90 kilometres south of present-day Baghdad. Its proximity to contemporary geopolitical hubris will not escape our notice. Genesis is not the only source of what supposedly happened there. Falvius

Josephus, the first-century authority on Jewish history fleshes the story out for us. He says:

> Now it was Nimrod who excited them [the Babylonians] to such an affront and contempt of God . . . He also gradually changed the government into tyranny – seeing no other way of turning men from the fear of God, but to bring them into a constant dependence upon his own power. He also said he would be revenged on God, if he should have a mind to drown the world again; for that he would build a tower too high for the waters to be able to reach! and that he would avenge himself on God for destroying their forefathers![4]

What we are seeing is that Nimrod stands for more than just worldly power. This is also a spiritual struggle of the 'Powers that Be'.[5] Here is Tolkien or Harry Potter writ first-millennium BC style. That's what you get to when you start looking at these old stories not necessarily as factual history but as living myth that echo down the corridors of time, patterning our present-day condition. The archetypal message of Nimrod is plain: pride leads to violence because it lives a lie that cuts us off from the fullness of relationship with others. Ecocide, the death of nature (as in homicide but applied to the environment), is the extension of that violence into nature. If we want to understand the scorched-earth consequences of a Saddam Hussein, a Hitler, a Pol Pot or even those who have lied and abused political power closer to home, it helps to consider such a framework of interpretation.

What happens next in Genesis is that God trumps Nimrod by the expedient of confusing the people's language. The big-name contractors all fell out with one another, the lawyers presumably got very rich, and the nations, Genesis tells us, scattered unto the ends of the Earth. As the psalmist later put it, the bottom-line is always the same: 'Except the Lord build the house, they labour in vain that build it.'[6] So much for the world's first bid at technofix. So much for the first failure to grasp the importance of lifestyle change in the face of hell's high water.

* * *

Babylon was at the centre of the ancient civilised world. In terms of psychohistory it is the archetypal city – a psychic template that patterns and helps organise subsequent human experience. But what was it that placed Babylon so much at the heart of earliest recorded human history? Here we must revert to a bit of historical realism through anthropology.

We need to start with human evolution and the probable fact that, for millions of years, our proto-human forebears had evolved as hunter gatherers in the savannahs and forests of Africa. As *Homo sapiens*, 'we' probably established as a species distinct from the other apes around 200,000 years ago. The earliest fully human fossils are some 130,000 years old, but scientists consider we'd have been on the go a bit longer than that. It is thought that we came out of Africa around 100,000 years ago.[7] As Africa's neighbour by land, the 'Fertile Crescent' was probably the first part of the wider world to be settled. It comprises the vast arc of alluvial soils watered by the rivers Nile, Jordan, Euphrates and Tigris, stretching from Egypt, through the Holy Land, to Iraq and on into Iran. In Biblical geography the Fertile Crescent was home to the Garden of Eden. As populations swelled, hunter-gathering gave way to the first settled agriculture, villages evolved into cities, and later historians came to speak of the Crescent having been the 'Cradle of Civilisation'.

That designation, however, reflects a certain bias in what is meant by 'civilisation'. Abundant literary and archaeological evidence suggests that violence was central to the organisation of the early Sumerian-speaking cities of Mesopotamia – a name that means 'between the rivers'. The Nimrod story in Genesis 10 is fascinating because it purports to document this emergence of highly centralised urban power hand in hand with a governing system of social domination.

The environmental consequences of such a way of life were inevitable. Genesis tells us, and the archaeology confirms it, that the cities of kings like Nimrod were built of bricks and mortar. This has ecological implications. Firing bricks and feeding the workforce required massive fuel and food imports from the surrounding rural hinterland. Over time, pressure on the land led to the first regional ecological catastrophe in documented human history. In 1936 when Leonard Woolley published the findings of his seven years' excavation of ancient cities such as Ur in the deserts of southern Iraq, he said:

> Only to those who have seen the Mesopotamian desert will the evocation of the ancient world seem well-nigh incredible, so complete is the contrast between past and present . . . It is yet more difficult to realise that the blank waste ever blossomed [and] bore fruit for the sustenance of a busy world. Why, if Ur was an empire's capital, if Sumer was once a vast granary, has the population dwindled to nothing, the very soil lost its virtue?[8]

Because these people were literate and kept accounting records, the answer to Woolley's question was found scratched in their own hand on tablets of clay. It makes for a fascinating detective story. Around 3,000 BCE – that is, 5,000 years ago – roughly equal amounts of wheat and barley had been grown. But wheat can withstand only about half a per cent of salt in the soil, while barley can tolerate up to 2%. Deforestation and soil erosion had led to the silting up of rivers which, together with irrigation, raised the water table. Instead of flushing through and eventually down to the sea, salts found naturally in rocks, soils and therefore in river water, concentrated in the topsoil as irrigation water evaporated. The consequence of such salination can be charted through time in the ratio of wheat to barley production. It relentlessly declined over the course of a millennium until, by 1,700 BCE, wheat had virtually disappeared.

Contemporary reports spoke of the soil sometimes turning white, presumably from salt precipitating out. In the centuries that followed the agricultural base in the south collapsed. The metropolis had to move north to the area around modern Baghdad, and there the same sorry saga was repeated all over again.[9] What had once been a resilient ecosystem – its soil bound in place and kept in good heart by rich plant and animal biodiversity – became a brittle ecosystem, easy to damage but hard to repair. Eventually with the moisture-retaining humus gone little more than desert sands remained. As such, the bricks with which early civilisation was built extracted a price higher than that of human slavery alone. Demand for firewood and food with which to feed the workforce crashed much of the ecosystem. The deserts over which we now squabble for oil are but the remnant of paradise lost, Babylon's fabled Hanging Gardens having been long since relegated from the top seven league of world wonders.

Bitumen was the mortar often used to bind the buildings of Babylon. In a peculiar irony the first examples of accidental death that are cited in the Bible are found in Genesis 14:10, where some of the fleeing kings of Sodom and Gomorrah fell into bitumen pits after battling their foes in the Dead Sea valley of Siddim. We generally miss this curiosity because the King James Version mistranslates them as 'slimepits'. The truth has more poetic justice. The sheiks and warlords of the Middle East spent their time thrashing one another, trashing the planet and drowning in oil even then just as they continue so to do today.

It all seems almost wearily inevitable, but must it be so? Here we might cue and cut to the controversial twentieth-century feminist archaeologist Marija Gimbutas. Gimbutas was Professor of European Archaeology at the University of California, Los Angeles, from 1964 to 1969. From her excavation of Neolithic sites in 'Old Europe' she claimed that ancient Mediterranean cultures, bordering onto western Mesopotamia between about 6,500 and 3,500 BCE, were peaceable and, if not specifically 'matriarchal', then at least expressing a high level of equality between the genders.[10] The overwhelming consensus amongst her peers is that this fascinating woman exceeded her evidence and ignored contrary points of view.[11] Yet, something in the strength of her ideas has given them wide popular influence especially amongst feminists. She claimed that many early societies showed little evidence of fortification, few skeletal signs of violent death, equality of men and women in burial, and a proliferation of 'goddess' figurines that suggested the veneration of feminine deities. However, as Bronze Age invaders swept in using the horse to extend their range of warfare, this Arcadia was forced to become militarised and, along with that, patriarchal. This, as she might have put it, was what paved the way for Nimrod's empire. According to Gimbutas's take on archaeology, settlements then became fortified, skeletal evidence pointed to violent mass death, men's graves were honoured more than those of women, and out went the gentle goddesses and in came the male war gods.

Comparative anthropology suggests that patriarchy and high levels of violence are widespread but not inevitable to human societies. Like Josephus tells us with Nimrod bringing people into 'a constant dependence upon his own power', these attributes appear to be acquired cultural characteristics. But once violence becomes normalised in a society it perpetuates its own reinforcing spiral. Unresolved and even celebrated as heroic, violence in its many manifestations enters the psyche of children and knocks on down through the generations. Until recently a statement like that might have been dismissed as speculative psychobabble. No longer is that the case. Recent advances in neurobiology now provide graphic evidence supporting the view that violence really does breed violence. Brain imaging shows that when a child is reared in a high-stress environment, the parts of its brain associated with empathy and head-heart communication between the hemispheres appear to show stunted growth. In an article in *Scientific American*, Martin Teicher, a Harvard professor of psychiatry, discusses this and concludes that:

> Society reaps what it sows in the way it nurtures its children. Stress sculpts the brain to exhibit various antisocial, though adaptive, behaviours. Whether it comes in the form of physical, emotional or sexual trauma or through exposure to warfare, famine or pestilence, stress can set off a ripple of hormonal changes that permanently wire a child's brain to cope with a malevolent world. Through this chain of events, violence and abuse pass from generation to generation as well as from one society to the next. Our stark conclusion is that we see the need to do much more to ensure that child abuse does not happen in the first place, because once these key brain alterations occur, there may be no going back.[12]

If, today, we Google the name of our Babylonian warlord, the first item to come up is a British military website. With no hint of shame it says: 'The Nimrod MRA4 is a maritime reconnaissance and attack aircraft.' Nimrods have been a backbone to British defence policy from the Cold War all the way to Iraq and Afghanistan. As such, the archetypal spirit of 'the first on earth to become a mighty warrior' rumbles on like a rusty chariot down millennial corridors of time. The reason why Gimbutas touched the nerve she did is that her narrative, archaeologically flimsy though it may be, speaks a life-giving psychological truth. Shortly we will see that ancient literary evidence lends it some interesting support. It matters because it is a narrative less perhaps of history, but of possibility. It counterpoints the grim inevitability of violence and Professor Teicher's thesis about its possibly permanent knock-on effects. Thankfully, there is indication that even serious personality disorders such as psychopathy can, with care and patience, be treated.[13] Just as stroke victims can be helped to overcome brain damage, so can some of those whose childhood social environments disposed their cognitive and emotional functioning towards violence. Like all warlords, Nimrod is an extreme example of the human condition, but one that reflects an archetypal pattern that carries consequences. Inasmuch as these consequences ultimately affect the environment, and so climate change, we must press our exploration still deeper.

* * *

It is not just Noah and Nimrod that make ancient literary connections between human behaviour and flooding. There are dozens and, some scholars would say, several hundred early texts that echo flood narratives. The most important of these are *The Epic of Gilgamesh* and writings of the Greek philosopher Plato.

Gilgamesh is a work that has long interested Biblical historians because it is roughly contemporary with, and very closely parallels to, the story of Noah's flood. It was scratched out on clay tablets some 5,000 years ago in Uruk in the southern part of that primordial city's modern namesake, Iraq.[14] It is generally considered to be the 'oldest book' in the world. As the stories in it were probably circulating orally for some 700 years before they were written down, we're working here with near enough a 6,000-year-old X-ray of the human psyche.

Like Nimrod, the founding king of Uruk, Gilgamesh is another archetypal despotic city-slicker. His description makes him sound like Saddam Hussein's brutal son, Uday. We're told that 'his arrogance has no bounds by day or night'. His appetite for conscripts to his armies meant 'no son is left with his father', and 'his lust leaves no virgin to her lover'.

The people pray for respite. They're answered by Aruru, the mother goddess of creation. She takes a pinch of clay and fashions Enkidu – Gilgamesh's alter-ego. We're told that Enkidu is his opposite and yet 'his equal . . . his own reflection . . . his second self, stormy heart for stormy heart.' Here, then, is the primal hunter-gatherer that lurks behind the suave exterior of every buttoned-down suit and tie! Not for him are the comforts of the city. Instead, Enkidu lives with the beasts of the hill. His body is rough, his hair long and matted, and he is 'innocent of mankind; he knew nothing of the cultivated land.'

At length these psychological two halves of the same man meet and wrestle one another. In spiritual mythology the gods or mortals wrestle to procure blessing, just as Jacob wrestles the angel to this end in Genesis 32. After a worthy struggle, Gilgamesh throws Enkidu, but he stands in awe at his courage. A friendship springs up – in other words, both are blessed. So begins their journey in psychodynamic synthesis – town and country; civilised and wild; head and heart; ego self and shadow self.

But Gilgamesh is not happy. Like Nimrod he yearns for recognition. He complains to Enkidu: 'I have not established my name stamped on brick as my destiny decreed.' He proposes to put this to rights. They must set out on a heroic adventure. They will go to the 'green mountain' – probably Mount Hermon at the source of the Jordan in Lebanon – to slay Humbaba, the demon that guards it, and bring down the sacred cedar trees. Like Nimrod, Gilgamesh sees himself as 'the man who can clamber to heaven.' He boastfully tells Enkidu, 'I will raise a monument to the gods' with the timber that they fell.

The elders of Uruk worry about their brash young sovereign. Like many young men in the first half of life he's getting ahead of himself. That's what the 'cork' does. It thinks it can outpace the river. The elders warn him: 'Gilgamesh, you are young, your courage carries you too far.' But the lads ignore them. Off they go adventuring. Heedless of morality and ecology alike, they slay Humbaba and then, 'Gilgamesh felled the trees of the forest and Enkidu cleared their roots as far as the banks of the Euphrates.' Here, then, is the first stage of creating the Iraqi desert: they not only take the timber; by extracting the roots, they prepare the land for ploughing.[15]

But it's not all going to go their way. On seeing such wanton destruction, Enlil, the 'Lord of the Open Field', curses the pair. Meanwhile, Ishtar, the Queen of Heaven, seems to have taken a shine to Gilgamesh and offers her hand in marriage. He bluntly turns her down. In effect, he rejects the potential wisdom incarnated by the goddess – the feminine face of God. To the ancients, wisdom, we should remember, meant 'the right application of knowledge'. In what maybe counts as the earliest documented trumpet blast of misogyny, Gilgamesh tells Ishtar that her love is like 'a castle which crushes the garrison, pitch that blackens the bearer, a leaky skin that wets the carrier'. Ishtar responds with the due fury of a woman scorned. She goes to her father and persuades him to set loose the Bull of Heaven to wreak vengeance. But the lads triumph over and kill that too! Ishtar is incandescent. She knows that Gilgamesh values above all else his comrade at arms – that part within him that takes a walk on the wild side. So it is that she curses Enkidu. He sickens and, without much ado, dies.

Thrown back onto his empty aloneness, Gilgamesh now wanders, deracinated and unsettled, over the land. He entertains himself killing lions, thus ancient images show him holding up their carcases like one of the last of the Great White Hunters that he is. Having seen what became of Enkidu, he now lives in fear of the prospect of his own death. Indeed, he's constitutionally unable to face the possibility of death. Such is what often happens to people who have sacrificed the inner life of the soul for outward egotism. Like Tolkein's Gollum in *The Lord of the Rings*, they wear spiritually thin and become spectres unto themselves. They may decry all religion as a 'cult' but then make cults out of themselves. As an example of such inflation, check out that self-proclaimed 'clear-thinking oasis', The Official Richard Dawkins Website – complete with its online store for self-promoting atheist t-shirts ($20) and even a choice of two Richard Dawkins car bumper stickers! Conrad put such vacuous hubris in perspective when he said of the colonial ivory trader Kurtz in *Heart of Darkness*, 'he was hollow at the core'. Similarly, Gilgamesh is fully on the slide into denial and nihilism. He tries desperately to compensate for his emptiness by searching obsessively for the secret of immortality. He's willing to travel unto the ends of the Earth to find it – as if immortality could ever be grasped and held onto by the transient ego self! So it is that, like Satan in the Book of Job, he spends his days in a kind of compulsive autism, wandering, as Job tells us, 'to and fro on the earth . . . walking up and down on it'.[16]

One day on his agitated travels Gilgamesh arrives at the home of Siduri – 'the woman of the vine, the maker of wine.' As we can see in Sufi poetry, in Christ's first miracle of the water into wine, and in the story of the Last Supper, the wine-bearer in Middle East spirituality symbolises 'reality revealed' through spiritual intoxication. Here, then, is a chance for Gilgamesh's redemption. The lovely Siduri sits there in her garden by the sea. Beside her is the golden bowl and the golden fermentation vats that the gods gave to serve her vocation. In what probably counts as the first-ever written-down speech of a woman, and surely one of the most elegant, she lifts her eyes to him and says:

> Gilgamesh, where are you hurrying to? You will never find that life for which you are looking. When the gods created man they allotted to him death, but life they retained in their own keeping. As for you, Gilgamesh, fill your belly with good things; day and night, night and day, dance and be merry, feast and rejoice. Let

> your clothes be fresh, bathe yourself in water, cherish the little child that holds your hand, and make your wife happy in your embrace; for this too is the lot of man.

But the hubris of fevered pride is now a full-raging madness. The inflated ego has lost all grounding in nourishment and context from the deep roots of the psyche. It has inverted to shadow – expired to darkness as surely as the sun must set. Gilgamesh is now just an ego split off from his deep Self. He not only 'has a complex'; he has become one. He has become a driven soul, tortured in its own self-made self-referential hell. Day and night the fear of death goads him. He must press onwards, onwards, as if to rid himself of a never-ending itch. He must, as he says of himself, 'wander still farther in the wilderness'. So it is that he rejects the tender Siduri, just as he had rejected the love of the goddess in her previous guise. Unable to give life from within himself he cannot receive its gifts from another.

Gilgamesh then learns of a sage, Utnapishtim the Faraway, who lives across the ocean. To him alone, of all men, the gods have granted the gift of everlasting life. It's time for Gilgamesh to place his final call at the Last Chance Saloon. His face now pinched and drawn from privation, he's ready to do anything, to go anywhere. And so he crosses the great ocean, locates the sage and asks him: 'How shall I find the life for which I am searching?'

But Utnapishtim's not going to play ball. With Buddha-like stoicism he replies, 'There is no permanence . . . [Life] is only the nymph of the dragon-fly who sheds her larva and sees the sun in his glory. From the days of old there is no permanence.'

After a vain bid to steal the sage's secret, Gilgamesh winds his weary way home to Uruk. There the river of life catches up and draws him under. He dies a mortal death and is laid by his people, as their duty requires, to a hero's rest. But before he had left the land of Faraway, Utnapishtim the Sage had told him a story. He told how it had come to pass that he had found immortal favour with the gods. The story went like this.

Long ago the gods were angered by the ways of humankind.

They held a council and said: 'The uproar of mankind is intolerable and sleep is no longer possible by reason of the babel.' They resolved to rid themselves of the pestilence and to do so by setting loose a deluge. But first they commanded Utnapishtim to build a boat and 'then take up into the boat the seed of all living creatures'. For six days and six nights it rained. The abyss rose up, the dams of the 'nether waters' were pulled out, and the dykes were cast down until 'the surface of the sea stretched as flat as a roof-top'.

Eventually Utnapishtim's boat came to rest against a mountain. He released first a dove, then a swallow, and finally a raven. When the raven saw that the waters had receded and that he was free to feed again, he crawed. On hearing this, Utnapishtim got out of the boat and made sacrifice and libations at the summit. As the sole survivor of humankind, and as the gods' faithful servant, they bestowed upon him the gift of immortality. And so the sage composed a poem. Like Siduri's speech, it possibly counts as the oldest written poem in the world since it comes from the oldest book. But what interests us here is its pertinence to consequences of climate change:

> Lay upon the sinner his sin,
> Lay upon the transgressor his transgression,
> Punish him a little when he breaks loose,
> Do not drive him too hard or he perishes;
> Would that a lion had ravaged mankind
> Rather than the flood,
> Would that a wolf had ravaged mankind
> Rather than the flood,
> Would that famine had wasted the world
> Rather than the flood,
> Would that pestilence had wasted mankind
> Rather than the flood.

* * *

We find the same motif of an ancient flood linked to human hubris at several points in the writings of Plato, who was born in Athens in 427 BCE. In *The Critias* he describes the ecological degradation of prehistoric Greece. Deforestation had destroyed the sponge-like capacity of the ground to retain rainfall. The rivers and springs therefore dried up until, 'You are left as with little islands with something rather like a skeleton of a body wasted by disease; the rich, soft soil has all run away leaving the land nothing but skin and bone . . . And the shrines which still survive at these former springs are proof of the truth of our present account of the country.'[17] Plato twice tells us that he is speaking of events that took place some 9,000 years before his own era. The most celebrated case that he describes is the story of lost island of Atlantis.

Just as Genesis 6 in the Bible tells us that 'the *sons of God* saw that the daughters of man were fair, and they took wives for themselves', so Plato, similarly, tells us that the Atlanteans were part human and part divine. Here we see the ancient idea that human nature is a melange of the worldly and the godly, and that both are needed for the right ordering of things. Such a pedigree, according to Plato, caused Zeus to have high expectations of the Atlanteans. They had achieved an advanced civilisation and wanted for nothing. But their human weaknesses got the upper hand over their better halves. Hubris took root and their arrogance corrupted them. In a scene that could serve as a splendid indictment of our contemporary world, Plato says that Zeus, 'summoned all the gods to his own most glorious abode, which stands at the centre of the universe and looks out over the whole realm of change':

> His reason, so the story goes, was this. For many generations, so long as the divine element in their nature survived, they [the Atlanteans] survived, they obeyed the laws and loved the divine to which they were akin. They retained a certain greatness of mind, and treated the vagaries of fortune and one another with wisdom and forbearance, as they reckoned that qualities of character were far more important than their present prosperity. So they bore the burden of their wealth and possessions lightly, and did not let their high standard of living intoxicate them or make them lose their self-control, but saw soberly and clearly that all these things flourish only on a soil of common goodwill and

> individual character, and if pursued too eagerly and overvalued destroy themselves and morality with them. So long as these principles and their divine nature remained unimpaired the prosperity which we have described continued to grow.
>
> But when the divine element in them became weakened . . . and their human traits became predominant, they ceased to be able to carry their prosperity with moderation. To the perceptive eye the depth of their degeneration was clear enough, but to those whose judgement of true happiness is defective they seemed, in their pursuit of unbridled ambition and power, to be at the height of their fame and fortune. And the god of gods, Zeus, who reigns by law, and whose eye can see such things, when he perceived the wretched state of this admirable stock, decided to punish them and reduce them to order by discipline.[18]

As we saw when discussing Francis Bacon's *New Atlantis*, Plato's *Critias* is an unfinished text, and this is as far as it goes. But in *The Timaeus,* Plato tells us that there were four great deluges in prehistory, and in one of these, Atlantis 'was swallowed up by the sea and vanished' after 'a single dreadful day and night' of 'earthquakes and floods of extraordinary violence'.[19] In another dialogue, *The Laws*, Plato says that since time immemorial, cities have come and gone due to floods, pestilences and other massive upheavals. Following each deluge only shepherds survive as 'mere scanty embers of humanity left unextinguished among their high peaks'. Flooding has periodically caused 'a total destruction of the cities situated in the lowlands and on the sea coast'. This left behind 'a vast territory of unoccupied land' characterised by 'frightful and widespread depopulation'.[20] Little wonder, then, Utnapishtim's poem had the refrain: anything . . . 'rather than the flood'. The consequences of such a scourge were total. And while these descriptions fit well with tsunamis triggered by undersea earthquakes, we will see that climate change taking place at this time may also have been a factor.

Plato goes on to tell us in *The Laws* that the surviving shepherd folk were unfamiliar 'with the tricks of town dwellers for overreaching and outdistancing one another and the rest of their devices for mutual infliction of mischief'. Gradually they resettled the devastated land. Their pastoral ways yielded to settled agriculture as they started to put up walls and fences. Leaders arose 'and thus made the patriarchal groups into an aristocracy, or possibly a monarchy'. Rival groups emerged, laws were made, states established and so history repeated itself all over again.

In yet another discourse, *The Statesman,* Plato explores the recurrent idea of a bygone Golden Age from which humankind periodically falls.[21] The notion had been around at least since the poet Hesiod in the seventh or eighth century BCE. It also comes up in the Hindu traditions of Asia, with the cosmology of the four *yugas* – the ages of gold, silver, bronze and iron – through which human history moves in descending spirals. The idea is that human ways have a built-in tendency towards degeneration. Things naturally drift from bad to worse until, one day, they get so bad that the gods send in the horses of the apocalypse for a right shake-up. To us they're natural disasters. To the collective psyche – the realm of the 'gods' – they're moral purges – a 'judgement'!

The most graceful articulation of the Golden Age comes from Empedocles of Acragas – a philosopher who lived in Sicily just before Plato's time, between about 492 and 432 BCE. He was a scientist, a physician, a democrat, an egalitarian, a pacifist, a minstrel, a champion of the poor, a knight in shining armour to young women in trouble getting their dowries together . . . and seemingly, in his colourful choice of attire, he was something of a dandy or, perhaps, a shaman.[22] Plato called him the 'gentle muse' because he was the last of the philosophers to express his thinking in verse. Whereas Heraclitus, who preceded him, had dismally said that strife is what drives the universe and its human affairs around in constant flux, Empedocles insisted that love counts too. Strife pulls apart the 'roots' of life – these being the four elements of fire, air, earth and water – but love brings them back together again. Reality is an elemental dance animated by both strife and love, thus, as one of his poetic fragments has it:

> One after another the roots prevail as the cycle goes around,
> Fading into one another and increasing as their appointed turn arrives.
> For they are just themselves, and by running through one another
> They become men and all the other kinds of creatures,
> Now being brought together by love into a single orderly arrangement,
> Now being borne asunder by the hostility of strife,
> Until they grow together as one and the totality is overcome.
> Thus, in that they have learnt to become one from many
> And turn into many again when the one is divided,
> In this sense they come to be and have an impermanent life;
> But in that they never cease from alternation,
> They are for ever unchanging in a cycle.[23]

Empedocles believed that in epochs when strife dominates, the male gods of war are venerated, blood sacrifice is practiced, and society relentlessly breaks down. But when love restores the four elements into synergy with the One, the goddess is worshipped, offerings are of myrrh, frankincense and honey, and human community is restored.[24]

* * *

Our selective tour of the ancients now brings us to a position where we see glimpses of a possible bygone social order not unlike that popularised by Marija Gimbutas. I suspect that the present-day value of thinkers like Empedocles, Plato and the writers of Genesis or Gilgamesh is to be found less in the accuracy of their histories than in the prescience of their vision. History tells us about our past, but vision reveals the future. For my money, it matters less whether a Golden Age might once have been a reality than what the seed of its idea can still promise to germinate within the human psyche. In this way too, the ancients can be vindicated as prophets before their time. We can all draw our own lessons from them. The ones that I choose are that hope lies in rebalancing the human with the divine; that 'strife' which stirs things up and gets things going must be tempered with love; and that the 'masculine' must be held in synergy with the 'feminine'. Furthermore, there is a close relationship between social and environmental collapse. We might sum it up with the formula:

Hubris = pride violence ecocide

Earlier I pointed to an irony in the ancient sources. I suggested that while early civilisations undoubtedly caused massive damage to their environments, it is unlikely that such damage could account for deluge narratives on the scale they described. Their moral critique may be spot on for our needs today, but their attribution of cause and effect may appear exaggerated. To have imagined that the whole world could have flooded as some of the ancient narratives suggest appears over the top. But is that judgement necessarily valid? The question is an interesting one because it links these early sources to the anticipated effects of modern climate change.

Scholars have long accepted that accounts of ancient deluges, because they are so frequent and consistent, are probably vestigial memories of actual events. Tsunamis caused by undersea earthquakes, subsidence of the land, or island explosions set off by volcanic activity are amongst the prime candidates. Plato's Atlantis myth perfectly fits such a tectonic category because he tells us that whatever happened, it was very sudden. But in addition to these usual explanations, it is instructive to enquire into what was happening to the world as a whole during the Neolithic and Bronze Age. We know that the ancients had a bad sense of chronology so we must be cautious. And yet, it intrigues me that the general ballpark of time that our literary sources make their claims about covers the past five to twelve thousand years. What makes this period so very interesting is that this was precisely when coastal regions all over the world really were being inundated. Sea levels were rising at a rate very similar to that which some scientists believe will affect our own and future generations, because of global warming.

The most probable cause in ancient times lay in those natural cyclical oscillations in the Earth's orbit around the sun, and therefore of solar intensity, that drives the comings and goings of great ice ages. It is hard for us to imagine today, but at the peak of the last ice age some 20,000 years ago, most of Scotland was a mile thick with ice. Glaciers advanced as far south as Birmingham. Large areas of northern Europe were under ice to a depth of 3 kilometres. Because so much water was tied up, sea levels fell to far below where they stand now. This rendered large parts of what are now the Irish Sea, the North Sea and the English Channel dry land or marsh. Early humankind could follow by foot the herds that they hunted as they migrated between England and what is now continental Europe.

Elsewhere in the world, the spread of human beings was aided by, for example, being able to reach Australia on rafts or simple boats, because the Torres Strait was so much narrower. As the ice age drew to a close and the glaciers melted, the sea level rose inexorably all over the world. The land bridge across the English Channel flooded only 8,000 years ago, causing Britain to commence its island destiny. In *The Revenge of Gaia*, James Lovelock puts the following question:

> The sea level was 120 metres lower than now, and land equal in area to the continent of Africa which is now below water was then above it . . . Imagine there was a civilization 12,000 years ago with cities on the coast . . . Who among them would have believed an early climate forecaster who claimed that soon they would be 120 metres beneath the ocean?

Lovelock goes on to compare the temperature changes then with those that the world is probably entering into now.

> The difference between the long-term average [world temperature] and the ice age, 12,000 years ago, is just over 3°C. The IPCC 2001 report suggests that [world temperatures] might rise a further 5°C during this century. This is about twice as much as the temperature change from the ice age to pre-industrial times . . . The changes likely in the world to come will, in their different ways, be as great or greater than this. True, the sea cannot rise

> more than another eighty metres, the amount of extra water which would be released if the ice of Greenland and Antarctica melted. But the world-wide torrid conditions would reduce the productivity of the remaining land and sea, and the loss of vegetation would slow the rate of removal of carbon dioxide and so sustain the hotter age for 100,000 years or more.[25]

Lovelock's scenarios are, it has to be said, on the outer edge of the IPCC's envelope of probability. Only their most pessimistic scenarios (as of 2007) anticipate the possibility of a temperature rise as high as 5°C this century. But Lovelock's historical point about the shock to our forebears is indisputable. Neither must we assume that sea levels would always have risen gently like the slow filling of a warm bath. The meltdown at the end of the last ice age – the Pleistocene – took place in spurts. It started about 20,000 years before present. In Scotland the Pleistocene is considered to have 'ended' some 10,000 years ago, thus ushering in the present interglacial epoch – the Holocene. But in the higher latitudes and altitudes of North America and northern Eurasia, the ice sheets persisted for longer. These continued to pour water into the world's ocean for a further 3,000 years. Indeed, it was not until this process was complete that the current geography of Britain, the English Channel, and all other coastal regions of the world assumed their present forms.[26]

At times the 'postglacial transgression' of rising waters could be very fast, giving rise to 'meltwater pulses'. During the first major such pulse, starting about 19,000 years ago and lasting less than 500 years, ocean levels rose by 10–15 metres. At 3 centimetres a year that is nearly ten times today's rate of rise.[27] More dramatic still, NASA scientist James Hansen claims that during another pulse 14,000 years ago – one that lasted 400 years – the sea level rose 20 metres; in other words 5 centimetres a year or a foot every six years. On this basis Hansen warns that if the Antarctic and Greenland ice sheets break up faster than the IPCC expects – and there is evidence of this happening due to the approach of various tipping points – then we could be looking at 5 metres of sea level rise this century. That's about ten times what the IPCC has thus far anticipated.[28] It has to be said that Hansen's opinions fall towards the more alarmist end of the scientific spectrum yet they remain within the credible ballpark. As this book went to press in the spring of 2008 I was given what appeared to be a reliable second-hand report that a model is being developed by a reputable scientific institute in Britain that corroborates Hansen's predictions. If this is the case, it has yet to achieve peer-reviewed publication and therefore cannot yet be counted as 'science'. But at this point I must draw back from speculation. Sometimes one just has to draw a line in one's investigations and, like Francis Bacon, concede that 'The rest was not perfected'.

Whatever our future might bring, the reality of the era towards which our ancient literary sources point was that meltwater pulses continued up until about 7,600 years ago. In low lying basins, flooding may sometimes have been catastrophically rapid. Once breached, sills or ridges of land could have completely given way under the force of rising waters dammed up behind them. In places such as at the mouth of the Black Sea and in what is now the English Channel, it is thought that plugs of ice may have suddenly broken loose like a champagne cork, resulting in a catastrophic rush of water inundating vast areas. Rapid incursions may also have taken place during the postglacial refilling of the lower Tigris–Euphrates basin in what now forms part of the Persian Gulf – close to the home of Gilgamesh. We so easily forget how much sea levels changed during early human history. Even the experts sometimes forget and have to remind themselves. For example, writing in the *Journal of the American Oriental Society* in 1979, Werner Nutzel reminded his fellow ancient historians that: 'About 14,000 BC the entire present Persian Gulf was a dry valley . . . From 14,000 to 3,000 BC the Gulf slowly began to fill itself up [to its present average depth of 50 metres] . . . To search for prehistorical Mesopotamian cultures, we must first always reconstruct the resulting changes . . .'[29]

As the Geodynamics Group from the Australian National University put it in expanding this theme in a 1995 research report: 'Excavations at Ur and elsewhere have led to evidence of a flooding event at about 4000–3000 BC and it is tempting to associate the Sumerian 'Flood' legend with the peak of the Holocene transgression during the flooding of the low-lying delta region when sea levels rose perhaps a few metres above present between about 6,000 and 3,000 years before present.'[30]

For Scotland, the ending of the ice age had very variable effects. Places that had been subjected to the highest densities of ice had been pressed down by its incredible weight into the semi-molten mantle of the Earth – a process known as 'isostatic equilibrium'. As the weight of the ice lifted, this land rose slowly back up in 'post-glacial rebound'. As such, many parts of Scotland rose in pace with the rise of sea level, and sometimes faster. Today this leaves raised beaches along many of our coastlines – narrow coastal ridges along which farmland and villages stretch, often with sea caves left high and dry in the cliff faces behind. Other parts of Scotland at other times in the early Holocene experienced massive flooding, especially during the so-called 'main postglacial transgression' that reached its maximum in Scotland around 6,500 years ago. At this point, a map of the territory would have looked very different to how it is now. The sea had invaded the Firth of Clyde and filled Loch Lomond. The Forth valley was flooded almost as far west as Aberfoyle. Whales ventured west beyond the present site of Stirling. Rising waters almost separated the Highlands from the Lowlands with a land bridge only 12 kilometres wide linking north and south. And the Hebrides and the Orkney islands, which may each have been one large island, became progressively fragmented as is evidenced, for example, by the partial submergence of chambered Neolithic cairns and forts in the Uists.[31]

As if steadily rising sea levels were not enough of a challenge to coastal communities, Mesolithic settlements in Scotland were overwhelmed by a tsunami around 7,000 years ago. It was caused by the 'second Storegga Slide' – a massive submarine landslide of glacial sediments down the continental slope off Norway. Comprising a sediment shelf the size of Iceland, this sent an enormous wave through the North Atlantic. It must have caused at least a ripple down west of the Pillars of Hercules where Plato located his Atlantis because, in Scotland, it deposited sediments 4 metres above normal levels and up to 80 kilometres inland.

Such are the realities on the ground by which ancient vestigial memories were perhaps shaped. Even today these dynamics are still with us. Most people are not aware that parts of Scotland are still slowly rising in the rebound of isostatic equilibrium from the ice age. As little as 2,000 years ago, the rate of rise remained sufficiently fast to cause earthquakes with magnitudes of 6 on the Richter scale – nearly ten times more powerful than the one that shook much of England in February 2008. In a paper published in *Scotland After the Ice Age*, Ballantyne and Dawson point out: 'Archaeologists seeking explanations for the destruction or abandonment of some early settlements in Scotland would be wise to consider the possible effects of such events.'[32]

Neither should it escape our notice that when large new hydro-electric dams are filled, the weight of the water often causes earthquakes. These have been recorded at some hundred sites around the world affecting even continental regions that were previously seismically stable. The most devastating to date measured 6.3 at the Shivaji Sagar Lake at Koyna, India. This claimed 200 lives, left 1,500 homeless and seismic activity has continued over the subsequent four decades.[33] Tsunamis and climate change are not usually presumed to be linked, but they could become so if flooding is rapid, causing the Earth's crust to become locally stressed. As such, it is not inconceivable that some of the flood-related tectonic activity to which Plato refers could have been triggered by the weight of water imposed on coastlines during the post-glacial transgression.

Again, we touch on speculative territory here, but what is not in question is that global warming will make coastal populations more vulnerable to tsunamis whatever their cause. This is because constantly rising waters mean that people will tend to find themselves living closer to the edge and therefore be more vulnerable than they might otherwise have chosen to be.

New Scientist in an editorial comment on James Hansen's projections points out that a sea-level rise of 10 metres would rob the Earth of 2% of its land surface. This agriculturally rich coastal and delta fringe is home to fully 10% of the world's population and it includes much of New York, London, Sydney, Vancouver, Mumbai, Shanghai and Tokyo, as well as vast swathes of Bangladesh.[34]

There is another potentially powerful card that could be played in trying to understand the environment of ancient humankind and its implications for us today. In a *Scientific American* article in 2005, paleoclimatologist William F. Ruddiman, an emeritus professor at the University of Virginia, describes his theory that global warming started well before the industrial era. He dates it back to the origins of deforestation and agriculture in Mesopotamia and China commencing 11,000 years ago. This, he speculates, may have saved the world from reverting quickly back into another ice age such as many scientists consider it might otherwise have been expected to do – thus the exhortation, common until recently, that we're 'due' another ice age.

By examining detail in the data from the Vostok ice core, Ruddiman observed that methane and CO_2 levels seemed to rise unexpectedly, starting about 8,000 years ago. They also appear to dip periodically after major human epidemics. At such times, some agricultural land temporarily reverts to forest, thereby re-capturing CO_2 and cutting the release of methane that would otherwise have bubbled up from paddy fields. Ruddiman concludes: 'As I see it, nature would have cooled the Earth's climate, but our ancestors kept it warm by discovering agriculture.' He cautions that this is not grounds for complacency about present-day climate change. Some, he says, might consider his findings to signify that global warming is a good thing (at least for those living in otherwise ice-bound northern latitudes). But 'others might counter that if so few humans with relatively primitive technologies were able to alter the course of climate so significantly, then we have reason to be concerned about the current rise of greenhouse gases to unparalleled concentrations at unprecedented rates'.[35]

For the subject matter of this chapter, Ruddiman's theory is salubrious. It invites speculation that the tail end of the post-glacial sea-level rise that shaped early human geography was accelerated by anthropogenic intervention. As such, the ancients would have been spot on, and without exaggerating, in supposing that their own environmental impact was contributing to some of the disasters they experienced! If Ruddiman is right, then in the absence of the actions of people like Gilgamesh and Enkidu we in Britain might still be able to follow the herds on foot to France. Just imagine Jeremy Clarkson hurtling along in his chariot to ogle all those skins elegantly shaped up on the Paris catwalks! However, Ruddiman's theory is one of many examples in climate science of a singular piece of work that, so far, lacks a broader range of corroborating data and scientific consensus to give it gravity. Indeed, some of his interpretations have, as would be expected, been contested. It's a debate worth watching if it develops further, but not yet one from which it would be responsible to conclude that early civilisations really did contribute to their own inundations.

It is sufficient for us to conclude that the ancients made a presumption that cataclysms of varying types were divine retribution for their waywardness. From a scientific point of view given present knowledge, that presumption was probably overstated. But in our own case, such a conclusion would almost certainly be understated. As we have seen, the main driver of contemporary climate change, according to the IPCC, is 'very likely' anthropogenic. This presents us with a strange irony. It is not so much the moral philosophers and religious prophets that are today calling us to eco-penance. It is the natural philosophers of hard science! The women and men in white coats from places like the Royal Society! And yet, the moral diagnosis – the links between patriarchal pride, violence and ecocide that we find so clearly expressed by the ancients – speaks to today's condition with a greater prophetic expediency than ever before.

As the Hebrew prophet Isaiah put it, writing a little earlier than Plato and Empedocles: 'The earth mourns and withers, the world fades and withers, the exalted of the people of the earth fade away. The earth is also polluted by its inhabitants, for they transgressed the laws, violated statutes, broke the everlasting covenant. Therefore a curse devours the earth, and those who live in it are held guilty.'[36]

The power of ancient thought is that, unblinded by the brilliance of self-styled 'Enlightenment', it constantly throws explanations for things back onto the human condition. For all that I love scientific method, its honesty and its explanatory power, I believe that it has allowed modern humankind to indulge too easily in the arrogance of forgetting that our own inner realities also shape the world. The metaphysical matters, for without it we miss the whole picture and have only ourselves to use as reference points. I don't want to kick out the science, but I would like to see our use of it tempered with some of the wisdom that a pre-modern world possessed.

In the next chapter I want to explore this interplay of outer forces and the inner life. To understand the modern condition we need to look at what shaped the psyche of Western societies. My thesis is that the sustained onslaught of hubris, and especially its intensification during the early modern era, diminished our collective capacity to sustain a rich inner life. This had implications that stimulated the rise of consumerism. But the story has a deep taproot, and since he speaks to it with a relevance that remains so contemporary, we will linger a little longer with Plato.

Chapter Six

DISSOCIATION OF SENSIBILITY

In his greatest work, *The Republic*, Plato depicts his mentor, Socrates, as hanging around with the privileged young men of the city discussing the meaning of life. Repeatedly they touch on a quality called *areté* (a-re-tay), usually translated as justice, virtue or excellence. It means the fulfilment of human potential – the fullest all-round expression of what a person is able to be. Socrates and his friends want to understand wherein virtues linked to *areté* reside. They agree to approach their discernment by analogy. They will examine such qualities first on the macro scale of the state, or republic, and then they will apply these principles of 'outer' life to the 'inner' dynamics of individual virtue.

Socrates kicks off by laying out his table. His ideal city state is the rustic idyll. In our sense today, what he describes is more of a village than a city. Men and women would spend their time in honest pastoral and craft work, living simply, caring for their children, guarding against poverty and war, eating a humble but wholesome vegetarian diet, drinking in moderation and, in their spare time, singing hymns to the gods. It's a bit like life back on a Hebridean croft but with olives and cheese instead of herring and potatoes; or an Amish community in North America, or a 'custom village' in Vanuatu where modern ways have been rejected by kinship groups that choose to retain tradition. Socrates warns that 'ambition and love of money are . . . something to be ashamed of', and the test of right livelihood is that people 'leave their children to live as they have done'. In this way he presages the definition of sustainable development that the United Nations' Brundtland Commission came up with two and a half millennia later: namely, 'Sustainable development is development that meets the needs of the present without compromising the ability of future generations to meet their own needs.'[1]

But the most petulant of the city Sloanes, a young man called Glaucon, is aghast at these suggestions. He tells Socrates that he and his fellows expect meat plentifully on the table, homes furnished with gold, ivory and art, perfumes and adornments for the wives and courtesans, clothes and shoes, entire classes of servants and nannies (both wet and dry) for the children. He says: 'If you had been founding a city of pigs, Socrates, this is just how you would have fattened them,' and he demands what he calls the 'ordinary dishes and dessert of modern life'. To this the master replies: 'Very well. . . I understand. We are considering, apparently, the making not of a city merely, but of a luxurious city. And perhaps there is no harm in doing so. From that kind, too, we shall soon learn, if we examine it, how justice and injustice arise in cities. I, for my part, think that the city I have described is the true one, what we may call the city of health. But if you wish, let us also inspect a city which is suffering from inflammation. . .'

And so, with his sword now suitably blooded, the Platonic Socrates turns to his famous question and answer method, ruthlessly teasing out the contradictions in Glaucon's aspirations.

> 'Then I dare say even the land which was sufficient to support the first population will be now insufficient and too small?'
>
> 'Yes,' he said.
>
> 'Then if we are to have enough for pasture and plough-land, we must take a slice from our neighbours' territory. And they will want to do the same to ours, if they also overpass the bounds of necessity and plunge into reckless pursuit of wealth?'

'Yes, that must happen, Socrates,' he said.

'Then shall we go to war at that point, Glaucon, or what will happen?'

'We shall go to war,' he said.

'And we need not say at present whether the effects of war are good or bad. Let us only notice that we have found the origin of war in those passions which are most responsible for all the evils that come upon cities and the men that dwell in them.'

'Certainly.'

'Then, my friend, our city will need to be still greater, and by no small amount either, but by a whole army. It will defend all the substance and wealth we have described, and will march out and fight the invaders.'

. . . 'Yes'. . .

'Then, Glaucon . . . with such natures as these, how are they to be prevented from behaving savagely towards one another and the other citizens?'

'By Zeus,' he said, 'that will not be easy.'[2]

So it is that the terrible ironies of hubris unfold. The short answer to Socrates' question is that the savagery he anticipates cannot be prevented. The very materialism by which anyone exceeds their fair share of the carrying capacity of natural or social systems has violence inevitably as its undercarriage – for the most part retracted and hidden out of sight. And, by Zeus, holding that in check is not easy! Easier perhaps to bite the metaphorical bullet, to fashion bullets, and to dispatch them in adventures of imperial overspill. Easier to play the victim blaming game of infantilising the vanquished while trumpeting the self-proclaimed heroics of your own colonising 'civilisation'. Psychology calls it the splitting off and projection of the shadow; the compartmentalisation of reality along perceived axes of evil. That at least keeps savagery focused outwards, away from the empire and onto the barbarians where it belongs!

Meanwhile the bread and circus lotus eating of polite metropolitan company gets ever more surreal as hubris further inflates the collective ego and cultivates the living of a web of lies. As Homer shows in the *Iliad*, you can maybe take Troy with a thousand ships launched by Helen's face, but not without a wasting frenzy of ever more demented violence. Such was the manic rage of Ajax the Great while camped outside the gates of Troy that he confused a herd of sheep for his rivals and spent the whole night slaughtering them! The next morning his shame forced him to fall upon his own sword. A red hyacinth grew from his blood, expressive of lament. Red poppies grew from the fields of Flanders. In the end, hubris hath no end, but lament.

The terrible reality is that a willingness to take by the sword slices its pound of flesh from the soul. For violence hollows out the capacity to have an inner life. It does so by desensitising the ability to feel and to relate to others beyond the formalised tenors of seemly conduct. As such, it opens up the gnawing emptiness of inauthenticity in human relations – the lie. This is the chasm into which the retail therapy of consumerism pours, and here are the roots of nihilism.

The hollowing out of the soul dumbs down the inner life not just for the dominant, who choose to trade the power of love for the love of power, but also for those who fall beneath their influence. The proud always fear the humble. They fear the wise and the free because they can't control them. They speak Truth that cuts sharper than the sword. This is why totalitarianism – political control over all aspects of life – is synonymous with fascism. The free and the wise – they are the same thing – threaten to rupture the artificial ego-controlled reality of domination systems and let in the flood. The flood of what? Of Truth, and of passion, of empathy, of beauty and authenticity. Indeed, of the human condition as it could potentially be; of all that holds out hope for the salvation of that self-same human condition.

The wise have this effect on the foolish because wise people know that there is more to being a human being than just the cork on the river – their own little ego selves. Whether one calls it God or whatever, they are connected with, or opening up to connection with, depths beyond ego. As such, they embrace the unconscious – the realm of dreams, visions and myth – rather than keeping it at arm's length. The narcissistic ego cannot do this at any great depth. It lives in denial of psychodynamic truth. The egotist and egotistic institutions or nations are therefore always vulnerable to embarrassment. This is what can make the embarrassment of power such an effective activist's tool. Embarrassment is but punctured ego. Just as Dorian Gray couldn't bear the portrait that reflected his inner truth, so all narcissists fear the child's cry that might expose their imperial nudity.

As such, outer power, where it is self-seeking and forceful rather than dedicated to service and discernment, is always tempted to govern the inner lives of others. It must stand above contradiction. It must find even spurious justification for its presumption of superiority. The colonised see more clearly than the coloniser the compartmentalised contradictions this creates. 'First they sent the missionaries who gave us Bibles, then they sent the mercenaries who took our land.' Inner colonisation therefore runs hand in hand with outer colonisation. It is as if the usurpers who twist religion to this purpose had never read the passages where Christ with the Devil on the mountain rejected the temptations of worldly power.

There are many examples in history and literature that illustrate these dynamics. Here just one will be sufficient to fulfil my aim, which is to deepen exploration of what has happened to our inner lives. The historical period that I consider most pivotal for us in the West is early modernity. In the case of Great Britain, I will take this as heralded by the accession of King James VI of Scotland to a united 'British' throne with England in 1603. James was invited to bring about this Union of the Crowns to ensure a Protestant succession after Elizabeth died childless. At the start of his consolidated reign, England held Ireland as its sole substantive colony. But by 1625, when James shuffled off this mortal coil, colonial footholds extended from India to the Caribbean and North America. The British Empire on which the sun would never set was up and running. British mores would shape the world. Even countries that did not fall under British rule felt the backwash. For example, it is striking how often the great Russian novelists attribute innovation in agriculture, technology and economics to 'English' ways, though they often feared for what the social consequences might be for Russian culture.

With our eye fixed to climate change, it will be instructive to view the making of early British modernity through the lens of William Shakespeare and his 'Scottish' play, *Macbeth*.

* * *

Macbeth was probably first performed in 1605. It was the same year as the Roman Catholic Guy Fawkes failed in his attempt to ignite two tonnes of explosive under James and the English parliament in the Gunpowder Plot. The first thing to notice is that Shakespeare's play is bound up with freak weather conditions. Macbeth is on his way back from suppressing a MacDonald-led uprising from the Scottish Hebrides. He meets three witches who are in the middle of raising storms to sink ships at sea. They predict that he will become king, and spurred on by his own 'vaulting ambition, which o'erleaps itself', Macbeth and his wife plot to murder King Duncan and ensure the prophecy comes to pass.

A psychological chill sets in as he contemplates the act. Macbeth's very perception of reality dulls. He remarks, 'Now o'er the one half-world | Nature seems dead.' On the night of the murder a storm outwith the bounds of living memory causes damage to buildings. Some feel an earthquake. Animals behave perversely. A small owl is seen to bring down a towering falcon. Duncan's horses break loose from their stables, 'as they would make war on mankind'. We're told: 'Thou seest the heavens, as troubled with man's act.' In other words, inversion of the social order has found perverse reflection in the natural order.[3]

But why such a strong emphasis from Shakespeare on the supernatural? The answer is probably that, as a playwright licensed by the Crown, the English bard wanted to please his new Scottish patron. Previously Shakespeare had seemed at ease with the supernatural fauna of folklore. His *A Midsummer Night's Dream* had appeared around 1595, and it sat comfortably as a comedy in the pre-modern world of faeries. Indeed, it inspired some of Britain's finest pre-Raphaelite influenced art such as Sir Joseph Noel Paton's paintings of Oberon and Titania that hang in the National Gallery of Scotland. But in *Macbeth,* Shakespeare takes a dark turn. Here, round the sinister cauldron of Hecat, Queen of the Witches, he describes 'Elves and fairies in a ring | Enchanting all that you put in'.[4] Why, then, have the faeries fallen to the status of witches?

Probably the Bard had done his homework. Back in 1597 while still in Scotland, King James had published *Daemonologie*, a treatise on witchcraft. This lumped the faeries in with the darkest doings of the supernatural realm. Scottish clergy, obsessed with witch hunts, had gone about stirring up hysteria to a point where, as historians put it, 'by 1590 any relationship between human and spirit, whether fairy or elf, could be seen only as evil.'[5] The Protestant Reformation which became state established in Scotland in 1560 had overturned the corrupted Catholic religious order and dismissed pre-Christian remnants of nature religion as 'papist superstitions'. As such, even the inhabitants of folklore were wrong-footed in the vicious sectarian divide of the times. Precisely this point was made by a contemporary of Shakespeare and King James, Bishop Richard Corbett. His poem, 'Farewell to the Fairies', has the oft-quoted verses:

Witness those rings and roundelays
Of theirs, which yet remain,
Were footed in Queen Mary's days
On many a grassy plain;
But since of late, Elizabeth,
And later, James came in,
They never danced on any heath
As when the time hath been.

By which we note the Fairies
Were of the old Profession.
Their songs were 'Ave Mary's',
Their dances were Procession.
But now, alas, they all are dead;
Or gone beyond the seas;
Or farther for Religion fled;
Or else they take their ease.

The good bishop of Oxford, and later of Norwich, lived between 1582 and 1635. He was a renowned practical joker and a great humorist. When country parishioners crowded too close around a confirmation ceremony he ordered them, 'Be off there or I'll be confirming you with my staff!' Another time he hailed a gent of venerable facial adornment: 'You, behind the beard!' The poetry of this quirky gardener's son was published in 1647 – posthumously – and perhaps so with good reason. As an Episcopalian like James – that is to say, as one who believed in Protestantism with bishops rather than the grassroots democracy of Puritan or Presbyterian governance – he mixed it mischievously for Catholics and Puritans alike:

Lament, lament old Abbeys
The Fairies' lost command!
They did but change Priests' babies,
 But some have changed your land.
And all your children, sprung from thence,
 Are now grown Puritans...

We will return to Bishop Corbett shortly because the psychology of faerie offers insights into the lacuna that violence creates in the soul. For now, let us stay with the storm-raising witches. As part of James's tidying up of his spiritual persona it had become necessary, while he was still in Scotland, to consummate a Protestant marriage. In 1589 he secured the hand of fourteen-year-old Anne of Denmark. But it was a year distinguished by a most spectacular flurry of freak storms. Anne's ship was initially driven back from Scotland and onto the coast of Norway. James himself went to fetch her and was again vexed by storms and fog.

Today, and with the benefit of scientific hindsight, we might see James's storms as probably having been linked to natural climate change.[6] Between the tenth and the fourteenth century, Britain had enjoyed a relatively warm and mild climate. But following this Medieval Climate Optimum, and continuing through until the mid nineteenth century, Europe as a whole suffered what is popularly known as the Little Ice Age. Due to both a cyclical dip in solar intensity and a spate of volcanic eruptions throwing reflective dust into the upper atmosphere, warm summers had stopped being dependable from about 1300. In some places and at some times, most notably during the Great Famine of 1315–17, Europe's livestock died, famines broke out and marginal farms were abandoned. Folklorists consider that this may be the source of tales like Hansel and Gretel, where parents abandon their children in the wolf-filled woods. From about 1550 onwards, Alpine glaciers spread and sometimes overwhelmed villages – notwithstanding the beseeching prayers of local clergy. Europe reached its coldest point around 1650 – the so-called First Climatic Minimum. The Thames had frozen over for the first time in 1607, and the following summer Lake Superior stayed frozen until June.

As James saw it, nature's providence had fallen under the spell of evil. Here was a Protestant realm besieged by the storms of spiritual treason. The finger of blame pointed to witches' covens presumed to be operating from both Copenhagen and North Berwick – a fishing village east of Edinburgh. Such was almost certainly the backdrop that inspired *Macbeth*'s weird sisters. Where Shakespeare wrote 'Though his bark [a small ship] cannot be lost, | Yet it shall be tempest-tossed', he most likely meant that James's own vessel on the mission to fetch Anne had only survived disaster thanks to the divine hand.

The storms of 1589 were therefore seen as spiritual maelstroms. At stake was God's plan for post-Reformation Scotland in a fragile emerging alliance of European Protestant states. The demonic spiritual politics of 1589 played themselves through again in the England of 1605 as the Gunpowder Plot would prove. The political paranoia of the era is plain from the preamble that can still be found printed inside most editions of the Authorised Version of the Bible. Commissioned by James and first published in 1611, it grandly announces its dedication as being to, 'The Most High and Mighty Prince James, by the Grace of God, King of Great Britain, France, and Ireland, Defender of the Faith'. What follows must rank amongst the most obsequious passages of political sycophancy ever written. It claims that had it not been for James's accession to Elizabeth's throne, 'some thick and palpable clouds of darkness would so have overshadowed this Land, that men should have been in doubt which way they were to walk'. It tells that the nation remains under danger of being 'traduced by Popish Persons at home or abroad', but thankfully, 'Your Majesty, as of the Sun in his strength, instantly dispelled those supposed and surmised mists, and gave unto all that were affected exceeding cause of comfort.'

* * *

But the comfort of some had been purchased at exceeding cause of discomfort to others. Following the 1589 storms the exigencies of spiritual war justified the most virulent response. Guantanamo Bay has nothing on this. With unrestrained vigour, several dozen accused witches were rounded up on both sides of the North Sea and put to test on evidence no stronger than what had been revealed by hearsay or by others under duress. Torture was justified on the grounds that only extreme pain could liberate truth from the Devil's grip. Sleep deprivation was standard, for example, by chaining suspects to the prison cell wall with a 'witches' bridle' that fitted round the head and tortured the tongue and cheeks with iron spikes. The king and his council authorised the tortures and James personally interrogated suspects, both before and after their being put to distress, at his Palace of Holyrood House – today a tourist honeypot where the Queen stays when visiting Edinburgh.

One of the most celebrated Scottish suspects was Agnes Sampson, a herbalist and midwife, known also as 'the Wise Wife of Keith'. All her bodily hair was shaved off in an effort to locate the Devil's Mark on her 'priuities', and she was 'thrawn' with a cord twisted excruciatingly round her head. After duly confessing her charges she was hanged and garrotted on Edinburgh's Castlehill in January 1591, 'thereafter her body to be burned in ashes and all her moveable goods to be escheat and inbrought to our sovereign lord's [i.e. King James's] use'.[7]

Others suffered more brutally. Euphemia Maclean (who had benefited from Agnes's herbalism in seeking pain relief in childbirth) was sentenced to be burnt 'quick' – which is to say, alive, without the mercy of first being strangled. And the schoolmaster Dr John 'Fian' Cunningham had his fingernails pulled up and needles thrust beneath. When this failed to extract confession, his legs were placed in the 'boots' – metal tubes, into which wedges were hammered until the marrow squeezed from the bones. *Newes from Scotland* – a report published in 1591 which is considered to be the earliest tract on Scottish witchcraft – gives detailed accounts of the tortures and says that the boots 'inflict the most severe and cruel pain in the world'.[8] The king and his council authorised these procedures in accordance with the law and the cost of Dr Fian's eventual execution was recorded in meticulous detail in the Edinburgh Burgh Treasurer's accounts as £5 18s 2d.[9] Like later Nazi executioners perpetrating their own versions of a 'final solution', the civil authorities evidently felt no need to hide their work. For them it required no apology. This was the prototypical war against terror on the axis of evil in defence of God's own realm.

* * *

In the Bible, the use of burning as a judicial punishment against an individual first comes up in Genesis 38. Here Judah accuses Tamar of whoring. He issues the command: 'Bring her forth, and let her be burnt!' But she manages to exonerate herself. She proves that he had been the unwitting father of the twins she carried!

Later in the Bible, Matthew's gospel goes out of its way to integrate Tamar and other 'fallen' women into the genealogy of Jesus.

Like Judah, King James practiced redemptive violence according to the old religious law. He was one type of Christian. To this day there are two.

* * *

In 1992 Edward Cowan, the professor of Scottish History at Glasgow University, wrote: 'It has been estimated that between three and four thousand persons were executed for witchcraft during the century 1590 to 1690. Scots often pride themselves on the bloodlessness of the Scottish Reformation, on the relative absence of martyrs on either side; these thousands were perhaps the true martyrs of the reformation era.'[10]

More recent scholarship for the Survey of Scottish Witchcraft at Edinburgh University identifies 3,837 individuals, most of them by name, who were accused of witchcraft between 1563 and 1736 while the Witchcraft Act was in force. Women accounted for 84% of the total. Up to two-thirds may have been executed.[11] Professor Brian Levack's recent book, *Witch Hunting in Scotland,* conservatively estimates 1,500 Scottish executions, though he cautions, 'of course we will never know the actual number'.[12]

Across Europe as a whole the 'Great Witch Hunt' started under papal authority and was continued by the Protestant reformers. It is thought to have killed 40,000–60,000 people. These figures, based on documentary evidence rather than folk memory, are much lower than earlier estimates including the much quoted exaggeration of 9 million that derived from a populist museum in America. But what probably cannot be overestimated is the effect that the terror must have had on the thought structures of far more than 9 million of our forebears.

Like Roman Catholic inquisitors who had preceded him, James sought to homogenise the mindset of his people. This was made utterly explicit in his policies to pacify the Western Isles – the area where Shakespeare pointedly located the uprising which Macbeth had suppressed. Royal writs from 1608 and 1616 spoke of the need to rectify 'scandalous reproaches' against the nation such as 'poor souls being ignorant of their own salvation . . . void of God's fear and our obedience'. It was deemed necessary to counter 'the continuance of the barbarity and incivility amongst the inhabitants'. To achieve such an aim, 'there is no measure more powerful to further his Majesty's principal regard and purpose than the establishing of schools in the particular parishes of this Kingdom where the youth may be taught at least to write and read, and be catechised and instructed in the grounds of religion'.[13]

Countless unwilling subjects of the British Empire would

come to experience similar policies: education that was for regimentation, not liberation. Intellectual freedom could be permitted within the logical rules of reason and holy writ, but not in ways that challenged metaphysical premises. As such, the heart implicitly had to be kept in constraint and so, something in the fullness of the human psyche had to wither.

This is what makes Bishop Corbett's *Farewell to the Fairies* so very interesting. Here we have a cultural artefact that touches on a gaiety – a sheer spontaneous *joie de vivre* – that seems to have been present in an older 'merrie' England, but was inimical to the prosaic concerns of the new world order. Did something change in England as the Bishop suggested? If so, was it anything of more than merely superstitious consequence?

To answer this question we have to consider what the realm of 'faerie' meant to the indigenous mind. In this, Scotland and Ireland can provide us with a clearer picture than can be gleaned through the Roman and Norman cultural overlays that fell on England. I suspect this is why so many English people roam north and west when seeking life's meaning. They're in search of something of their own that has been driven to the hills and winding byways.

The great Gaelic scholar John MacInnes says that in the Celtic mind faerie is 'a metaphor for the imagination'.[14] We see this in the recurrent notion that faerie is the inner source of poetry, music, art and myth. Faerie stands for the imagination as being something more than just the imaginary and unreal. It is the mind's capacity for 'liminal' experience – for ways of seeing and being that traverses inner 'thresholds', expands consciousness and so transforms reality. Here is spiritual perception expressed as that which penetrates to the interiority of things, for spirituality *is* the interiority of outward material and institutional forms. This opens to the eye of the heart a world that is personified, animated, alive and, so, 'magical'. We then become aware of a world replete with intrinsic value – the value that things have in themselves – that triggers awe and constellates respect. It brings our own lives into unison with the unfathomable river of nature. As the Scots mountaineer Nan Shepherd puts it in her enchanting book *The Living*

Mountain, the pilgrimage to a mountain 'is a journey into Being; for as I penetrate more deeply into the mountain's life, I penetrate also into my own.'[15]

Understood as a state of consciousness and being, faerie and its realm within the hollow hill becomes a metaphor for our own long-buried indigenous green consciousness.[16] Faerie quickens the jaded spirit and reinspires it with imaginary possibility. These things permeate the psyche of our culture far more deeply than most moderns realise. Consider the fairground carousel, the merry-go-round. The very name, 'fair', is cognate with faerie and the 'fair day' festivals or holidays – holy days – that divide the Celtic year into its quarters. Many a child's first visceral encounter with the mythical is when it rides a unicorn or some other fantastical beast – perhaps a Harley-Davidson motorcycle! – on one of these pulsating music-filled rings of vivacity – these transient urban faerie hills popped out like autumn mushrooms, dipping rhythmically up and down, overworld and underworld, high road and the shamanic low road – whenever the wild traveller folks of no fixed abode set up their 'circus', their magic circle, amongst the world-weary settled folks of town.

Such indigenous 'deep ecology' – such a synthesis of inner life and outer nature – is invisible to eyes abased by their own abuse. As such, men like King James experience only cynicism or paranoia when confronted by it. W.B. Yeats lays out the problem as an indictment in *The Celtic Twilight* where, with wry humour, he says:

> In Scotland you are too theological, too gloomy . . . You have burnt all the witches [and] discovered the faeries to be pagan and wicked . . . Carolan slept upon a faery rath. Ever after their tunes ran in his head, and made him the great musician he was . . . For their gay and graceful doings you must go to Ireland; for their deeds of terror to Scotland.[17]

And we duck the point if we read this literally. Yeats is not urging a Peter Pan-like belief in Tinkerbell dancing skimpy-clad down a leafy lane. That is only a projected image of inner archetype. Yeats is saying something much more radical. Yeats is telling us to get real about the living roots of creativity. He wants us to heal from our dried-up miserableness.

T.S. Eliot brings us to the same point by a different route. It is interesting that Eliot's own English ancestor, the Rev Andrew Eliot, had suffered 'great mental affliction . . . in the residue of life' because he had officiated in the witch trials at Salem, Massachusetts, 1692–93.[18] The bard's deeply important essay, *The Metaphysical Poets,* published in the *Times Literary Supplement* in 1921, puts forward his theory of the 'dissociation of sensibility'. He means by this the breaking up of the ability to feel and relate to life, thus:

> In the seventeenth century a dissociation of sensibility set in, from which we have never recovered . . . It is something which had happened to the mind of England between the time of Donne . . . and Tennyson and Browning; it is the difference between the intellectual poet and the reflective poet. Tennyson and Browning are poets, and they think; but they do not feel their thought as immediately as the odour of a rose. A thought to Donne was an experience; it modified his sensibility [and] these experiences are always forming new wholes . . . While the language became more refined, the feeling became more crude . . . [in some cases exposing] a dazzling disregard of the soul.[19]

In Scotland, the Hebridean poet Iain Crichton Smith endorsed T.S. Eliot's appraisal and added that 'some irretrievable damage was done to the Scottish poetic psyche . . . as far back as the Reformation'.[20] Smith draws this into his lament for the decline in what he calls 'the feeling intelligence'. We have reached a point, he says, where we are expected to explain everything, as if life is a set of questions and answers from the Shorter Catechism. We will 'not allow the intelligence to be other than that of the relentless logician' and this is 'because the feelings have been lost, it is because *we are afraid of our feelings, and we have substituted a dead intelligence in their place*'. When that happens, he says, the culture loses touch with the 'feminine' and so becomes bereft of tenderness.[21] The logical positivism described earlier had come of age: '. . . if you can't kick it, you can't count it.'

This, then, is how hubristic violence destroys the inner life. It strips sensibility not just from the violent but also from those who suffer violence. Its lie is to represent the inner life as a lie. This is why I see tackling the roots of violence and untruth as being even more important than dealing with the immediate and overt symptoms expressed as climate change. The two must, of course, be taken in tandem, but violence to both the material world and truth is the more fundamental issue.

As it is, we are left with a society that may be highly individualistic in an egocentric sense, but where people are not, to use the word as Jung does, 'individuated' – able to stand back from the herd instinct of the crowd and appraise life from a more deeply constellated grounding.[22] We are left, in the words of Thomas Gray, as a society of those whose 'sober wishes never learn'd to stray . . . far from the madding crowd's ignoble strife'.[23] In his 1925 poem 'The Hollow Men', Eliot expressed this even more cuttingly. Our nemesis as 'lost violent souls', he suggested, is to become mere 'hollow men'. We whisper in vacuous conformity like rattling straws in the wind until the world ends. And it ends not in any dramatic manner such as Guy Fawkes had plotted for King James in 1605. It ends, as the poem alludes in an explicit resonance with the Gunpowder Plot, 'Not with a bang but a whimper'.[24] As such, the ending of the world is banal. It is a slow holocaust. And as Hannah Arendt would show in her celebrated study of Nazi atrocities, banality is the essence of evil – it only takes the good to do nothing for it to flourish.

Applied to global warming all this would be so very depressing were it not for the capacity of the human spirit to heal and to be re-infused with life. Here it is so very important to counterpoint grim reality with the magic we have been exploring as 'faerie'. Perhaps the reason why this was historically so repressed in Scotland is that it was so strong in the popular culture. Professor G. Gregory Smith's 1919 study, *Scottish Literature*, remains one of the most influential books of Scottish literary criticism to this day. He maintains that 'two moods' interweave in the national psyche – 'the Scottish antithesis of the *real* and the *fantastic'*. Either can 'invade the other without warning'. As such, reality and fantasy are 'warp and woof' as 'polar twins of the Scottish Muse'. Reality comprises the rational outer realm of the commonplace, the practical, and of catechisms and curricula. Fantasy is the inner world of imagination, meaning and magic, where the psyche, Smith lyrically says, is 'thrown topsy-turvy, in the horns of elfland and the voices of the mountains'.

This 'strange union of opposites', held in mutually fecundating tension, is Professor Smith's famous, if virtually unpronounceable, 'Caledonian antisyzygy'. Let that be my gift to Scrabble players! Many consider it the most distinctive quality of Scottish literature. It surfaces in such cultural archetypes as Dr Jekyll (practical) and Mr Hyde (fantastic). It plays out in the music too. Neither is Smith circumspect about the crazy wildness of it all. After all, he says, none other than Robert Burns himself is our constant reminder that the fantastical is animated 'by fairies, and brownies, and witches, and warlocks, and spunkies, and kelpies'.[25]

Again, we are not literally playing Tinkerbell here. We are negotiating liminality and that is the paradoxical magic of it. The poet Hugh MacDiarmid underscores our point. He writes of Caledonian antisyzygy as 'Scotland's distinctive function in the world' – one that is connected to 'Freedom – the free development of human consciousness'. But he takes care to quote, approvingly, from an unnamed critic who said that antisyzygy 'partakes not of fairyland, but of the enchantment of life itself – indescribable as a sea of changing colours touching all the shores of possibility'.[26] But how very interesting is that choice of words! Even while taking wise practical distance from kitsch articulations of faerie, the writer embraces the fantasy realm of enchantment! How very antisyzygic! How exemplary of the magical realism of the native Celtic mind!

And England? How revealing that Kipling's children should have summoned Puck unwittingly from out of Pook's Hill – the archetypal faerie hill of the cultural collective unconscious. Puck tells them that he came to Old England 'with Oak, Ash and Thorn', and only with them would he ever leave. But he is now the last faerie in the land; the last, because the rest 'couldn't get on with the English for one reason or another'.[27]

One of the children remembers a verse from Bishop Corbett's 'Farewell', the one that describes the *joie de vivre* lost with Elizabeth and James. He dejectedly tells Puck, 'When I was little *it always made me feel unhappy in my inside'*.

And there we see the cultural lacuna. There we see the hole inside that festers, hurting, lonely, like a part-forgotten yet wholly unrequited love . . . and it's there not just in English folks, but in so many of us – perhaps all of us who have been shaped by the modern condition.

'I read the news today,' sung The Beatles in 1967 in *Sergeant Pepper's* – one of the top ten best-selling albums of all time. 'And though the holes were rather small | They had to count them all | Now they know how many holes | It takes to fill the Albert Hall.'

For like concert-goers seeking their fill, *we are those holes in the road*. We who suffer from varying degrees of dissociation in our sensibility. We, the lonely hearts of Sergeant Pepper's Lonely Hearts' Club Band.

* * *

And what was Puck's cure for the children's unhappiness inside?

It is simple. He tells them stories; their history, albeit a somewhat rumbustious colonial version as was Kipling's Victorian wont.

But he opens their inner eyes and activates the *Mythos* – the deep structures of identity.

From out the faerie hill he restores something of the flow of magic set in place when time began.

Really, when you think of the *Sergeant Pepper's* album sleeve with such mind-blowing luminaries as Albert Einstein, Aldous Huxley, Carl Jung, Dr David Livingstone, Sri Paramhansa Yogananda and Marilyn Monroe, Puck must be hidden there . . . somewhere in the flowerbeds at the bottom.

Indeed, the album is riddled through with fairground allusions including organs played 'for the benefit of Mr Kite' –

originally a performer in Pablo Fanque's nineteenth-century circus, a picture of whom had inspired John Lennon. And Lennon reinforces the sense of faerie in the *Anthology 2* alternative version of *Sergeant Pepper's* last track, 'A Day in the Life', which starts with the bizarre count-in of 'Sugar-plum fairy, sugar-plum fairy'.

How ridiculous! But it's as if something is bursting to get out that's been long held in. From where else could Paul McCartney have got the controversial refrain to that track's carnivalesque crescendo, 'I'd love to turn you on'! That's precisely what so much of 'faerie' is all about, deep in the traditions of Celtic and other peoples.[28] These things do not come from nowhere in the cultural psyche. They're part of our make-up. The course of history has merely turned off or, at least, diminished something wondrous inside us all. Today that fantastical creativity yearns to turn back on, for it is a living creativity; a 'quickening' from within to nourish the spirit and feed the roots of the world.

* * *

In this chapter we have explored the way that culturally embedded violence damages people's capacity to have an inner life. It establishes norms that deaden the soul and create a pliant, hollow populace.

Next I want to look at how Western capitalism was able to exploit this evisceration and thereby manufacture the waste land of consumerism. But before going there, let me acknowledge that this has been a harrowing chapter to write. I'm sure it will not have been comfortable for many of my readers either. There's a line in one of Eliot's essays where he talks of how one can be 'oppressed by the burden which he must bring to birth'.[29] I remember feeling just that way while giving a lecture about violence in New York State in 2003. Let me share what happened because it has a lovely affirming outcome.

It was just after the start of the second Gulf War and I was acutely aware how uncomfortable many in the audience felt at that time about being American. They felt trapped as members of an oppressor's camp that they couldn't own their membership of. It's the same for most of us in many walks of life today. We, too, are children of the consumerist world that drives climate change. And yet, if we can't get to terms with and own our complicity, it will blind us from being able to see and tackle the issues.

At the end of my talk a man in the audience came forward. He told me the following story that kind of put things into a wider perspective.

'You know,' he said, 'I used to be a hippie back in the days when hippies were real hippies who set out to change the world. And a couple of my good hippie friends travelled to Cuba, because they wanted to shake hands with Che Guevara.

'Well, they managed to get an audience with the great Che, and of course, being real hippies, they started off by apologising for being Americans.

'And do you know what?' he continued. 'The great Che, living out there on Cuba . . . so far away from where it all happens . . . he just smiled back. And he told them: "Don't you apologise for being Americans. You're the lucky ones. You see, you get to live in the belly of the beast."'

Chapter Seven

COLONISED BY DEATH

So what is the belly of the beast? What are the juices by which consumerism digests the energies of human life and sets loose environmental symptoms of which climate change is but the most pressing? To answer these questions we must build on the historical perspectives that we have just examined. We need to set them as the backdrop to the rise of modern commerce and, especially, advanced marketing techniques that force the pace of consumption. Let us be clear that our concern is not with marketing as the provision of information. Neither is our quibble with economics as the means of supplying the basic necessities of life. Our concern is with what economists call 'the diminishing marginal return of utility'. This is the principle by which each added unit of 'utility', or satisfaction, requires ever more input of 'stuff'. For example, the first pack of butter gives much utility, the second one a little bit less, and so on until, by the time that a family might have many packs of butter, the addition of yet another adds very little extra pleasure. Our concern is therefore with what happens when entire affluent societies find that the cutting edge of added satisfaction has shifted far along this 'utility function'. As a result, the achievement of just one more unit of supposed happiness demands many units of material input and these, at a correspondingly escalating social and environmental cost.

Marketing as we know it today is largely a phenomenon of the twentieth century. It was made possible by technologies that facilitated both mass media communications and surplus production. Up until the First World War it was generally presumed that the function of economics and its application through industry and commerce was to satisfy human needs. But from the 1920s, and especially after the 1940s, a new way of thinking emerged – the explicit creation of wants.

Writing exactly fifty years ago in his acclaimed book, *The Affluent Society,* John Kenneth Galbraith used these sombre terms:

> The general conclusion of these pages is of such importance for this essay that it had perhaps best be put with some formality. As a society becomes increasingly affluent, wants are increasingly created by the process by which they are satisfied . . . Increases in consumption, the counterpart of increases in production, act by suggestion or emulation to create wants . . . It will be convenient to call it the Dependence Effect.[1]

More recent writers like Clive Hamilton and Richard Denniss have named it 'affluenza'. As they see it, rich societies: '. . . seem to be in the grip of a collective psychological disorder . . . We have grown fat but we persist in the belief that we are thin and must consume more . . . Affluenza describes a condition in which we are confused about what it takes to live a worthwhile life. Part of this confusion is a failure to distinguish between what we want and what we need.'[2]

That confusion is not accidental. Human beings who know what makes for a worthwhile life make bad consumers. Such people are relatively self-contained. Beyond seeking a comfortable level of dignified sufficiency – food, warmth, shelter, recreation – they don't need to keep buying happiness. The only reason that consumerism has taken off the way it has in the West and wherever Western mores have spread is because the modern mind, or the 'postmodern condition' as some would see it, is insecure and ungrounded. Its people generally need socially sanctioned external reference points to feel good about themselves. We all do, but it's a question of degree and proportionality. Part of this need can be attributed to lack of personal growth. Most of us go through the stage of needing to have more and more 'things' during what is, psychologically speaking, the first half of life – the stage of ego formation. But maturity in the second half of life – the stage of deep grounding – should see a mellowing out.

We see this in the kind of older people who'll say, 'We've got all we need now.' Often they are not outwardly rich. Their wealth is inner, and as such, it keeps growing. Their riches reside in relationships and in their capacity to appreciate what they've already got. They draw their 'utility', to use that awful economist's word, from activities like home baking, gardening, making and mending things, education, the arts, making love, telling stories to children and playing a role in the community. Provided they don't spoil it all by flying off to see the grandchildren in New Zealand twice a year, their impact on the environment can be relatively light.

To live a life that deepens into frugal but fulfilling sufficiency is not what creates problems in the world. As we heard Socrates tell, that is how people can 'leave their children to live as they have done'. The problem of consumerism comes about when the competitive pursuit of utility trashes the very means by which satisfaction might be derived. This is consumerism not as a passing stage of life, but as life retarded. It is consumerism resulting from the evisceration of inner life by powerful destructive forces – Leonard Cohen's 'blizzard of the world' that has 'overturned the order of the soul'. This is what perverts the social order and, as *Macbeth* symbolises for us, inverts the order of nature in so doing.

Global warming is but one of the more pressing symptoms. War, resource depletion and poverty are others. Our feelings of despair derive from the seemingly hopeless inevitability of it all. Our insistence on the rediscovery of hope must therefore come from understanding the causes, seeking out ways of addressing them, and never giving up on the potential that might be emergent within the human condition. What is so fascinating looking back over the twentieth century is the extent to which what became of us was so very understandable. It was so, as we shall now see, precisely because it was so carefully engineered.

* * *

In May 1925 the Associated Advertising Clubs of the World held its annual conference in Manhattan. They were addressed by Herbert Hoover (of Agricola translation fame), now Secretary of Commerce and, in another four years' time, destined to become President of the United States. Echoing the spirit of the times he told them: 'The older economists taught the essential influences of "wish", "want" and "desire" as motive forces in economic progress. *You have taken over the job of creating desire.* You have still another job – creating goodwill in order to make desire stand hitched.'[3]

The conference resolution that year was 'War is the foe of trade'. It would have resonated with Hoover's pacifist Quaker upbringing even if 'creating desire' for material things might have sat less easily. But these were exceptional times. Fifteen million people had just been killed in the First World War. Psychologists were trying to understand the mass psychosis that drives war. World leaders were anxious to redirect human energies towards peace. War had shown that propaganda could capture men's hearts and drive them to death. Would democracy not be served better by healthy business competition than lethal military aggression? Creating a demanding common purpose around the pursuit of prosperity in the wake of wartime privation seemed a sensible way forward. It would please electorates. It would reassure corporations that feared market collapse in the absence of war's spoils. Democracy, psychology and commerce could profitably march hand in hand.

One of Herbert Hoover's advisers was Edward Bernays, a nephew of the Austrian psychotherapist Sigmund Freud. During the war Bernays had worked for the US government's Committee on Public Information. In 1928 his insights found expression in an influential book, *Propaganda*. A later work by this 'Father of Propaganda' had the shameless title, *The Engineering of Consent*. In his life – and it was an innings that ran from 1891 to 1995 – Bernays would go on to provide 'public relations' advice to several US Presidents and a plethora of corporations. *Life* magazine listed him as one of the 100 most influential Americans of the twentieth century. He proclaimed democracy, but his core message was authoritarian. The opening paragraph of *Propaganda* sets the tenor: 'The conscious and intelligent manipulation of the organised habits and opinions of the masses is an important element in democratic society. Those who manipulate this unseen mechanism of society constitute an invisible government which is the true ruling power of our country.'[4]

Central to his approach was the rhetorical question: 'If we understand the mechanism and motives of the group mind, is it not possible to control and regiment the masses according to our will without their knowing about it?' Big business needed to heed this question just as much as government. The world was in flux. The scale and pace of industry was changing. Today, as Bernays put it, 'mass production is profitable only if its rhythm can be maintained – that is, if it can continue to sell its product in steady or increasing quantity'. In days gone by, production had been artisan, bespoke. As such, he tells us, 'a century ago, demand created the supply; today supply must actively seek to create its corresponding demand'. But this puts the system at risk of becoming a victim of its own success. It gives consumers too much choice. Life is too short for them to make informed decisions. It follows, Bernays says, that 'society consents to have its choice narrowed to ideas and objects brought to its attention through propaganda of all kinds. There is consequently a vast and continuous effort going on to capture our minds.'[5]

Propaganda – a.k.a. public relations or 'spin' – is therefore crucial to marketing. This means getting to know consumers better than they think they know themselves. 'The successful propagandist must understand the true motives and not be content to accept the reasons which men give for what they do.' The 'old propagandist' of a passing age tried only to achieve a simple stimulus-response reaction in the customer, for example: 'Eat bacon because it is cheap, because it is good, because it gives you reserve energy.' In contrast to the old way, Bernays suggests, 'the new salesman will then suggest to physicians to say publicly that it is wholesome to eat bacon. He knows as a mathematical certainty that large numbers of persons will follow the advice.'[6] What sells is not so much the 'generic product' – bog standard bacon – as the psychological add-ons that go towards creating what marketers call the 'augmented product'. The fact that truth may become a casualty in the process rarely seems to have bothered Eddie Bernays. The fusion of mendacity and self-righteousness in his writing is quite stunning. The bottom line for him and his clients was that his techniques worked. Soon they were normalised on a competitive business playing field that increasingly spanned the world.[7]

By the late 1940s the world had seen another war. American corporations were again concerned with propping up their markets as peace broke out. Although much of Europe was still under rationing, mechanised production thrust Americans into an era of unparalleled satiation of their basic needs. Once again, corporations turned to the psychologists to keep ahead in the game. The iconic study of the times is Vance Packard's *The Hidden Persuaders*. Packard went round and interviewed industry movers and shakers in the late 1940s and early '50s before a sense of decorum caused them to become more guarded. His book cites the front page of *The Wall Street Journal* as saying: 'The business man's hunt for sales boosters is leading him into a strange wilderness; the subconscious mind'. Packard explains:

> Ad men began talking about the different levels of human consciousness. As they saw it there were three main levels of interest to them. The first level is the conscious, rational level, where people know what is going on, and are able to tell why. The second and lower level is called, variously, preconscious and subconscious but involves that area where a person may know in a vague way what is going on within his own feelings, sensations, and attitudes but would not be willing to tell why. This is the level of prejudices, assumptions, fears, emotional promptings, and so on. Finally, the third level is where we not only are not aware of our true attitudes and feelings but would not discuss them if we could. Exploring our attitudes towards products at these second and third levels became known as the new science of motivational analysis or research, or just plain M.R.[8]

Pushing this along was a new kid on the block. What Eddie Bernays had been to the advertising industry after the First World War, Dr Ernest Dichter was to the Second. Dichter was only the most prominent amongst a loosely knit school known as the 'depth boys'. Scouring the works of depth psychologists, especially Freud, Jung and Adler, these men set about trying to convert therapeutic insight to commercial gain.

Depth psychology had emerged during the late nineteenth and early twentieth century. It understood the human mind to have hidden layers that comprised the unconscious. The Age of Reason leading to the Enlightenment of the eighteenth century had, as we have seen, built on the ideas of philosophers like Francis Bacon, Lord Chancellor to King James. Reason was thought to be capable of explaining all things; penetrating all recesses of nature and the mind. Science should be 'value free' and avoid 'emotive arguments'. But all this sidestepped the human heart. It made people psychologically ill because it treated them in a mechanistic and utilitarian manner. Depth psychology therefore emerged as a reaction in the late nineteenth century.

Freud himself took the view that the very process of civilisation had created neurotic malcontents. He maintained that civilised norms channel the libido – psychological energy – away from the 'pleasure principle' of *Eros* and into the 'reality principle' that we might equate with *Logos*. The consequent neurosis is the psyche's revolt. Psychosomatic illnesses, phobias, eating disorders and depression are all the soul's way of signalling, as Shakespeare put it, that 'something is rotten in the State of Denmark'. Carl Jung went further. As he saw it, Freud was too obsessed with sex. The psyche also rests in a framework of *Mythos* – the collective unconscious realm of myth, archetype and all the 'numinous' or spiritual underpinning of life that comprises the metaphysical. Denial of this was at the heart of meaninglessness – the greatest disease of our times. The churches were to blame even more than the logical positivists and state. They had colluded in the ossification of their own domain – the flow of life.

In a childhood vision Jung had watched God sit on His

golden throne and lay an almighty cosmic turd. It dropped all the way from Heaven and smashed the cathedral to smithereens![9] Without spiritual awakening, without a laxative to the post-medieval legacy of spiritual constipation, Western humankind, like Plato's Atlantis, was in peril from its own powers. As Jung wrote just before his death in 1961:

> Modern man does not understand how much his 'rationalism' (which has destroyed his capacity to respond to numinous symbols and ideas) has put him at the mercy of the psychic 'underworld'. He has freed himself from 'superstition' (or so he believes), but in the process he has lost his spiritual values to a positively dangerous degree. His moral and spiritual tradition has disintegrated, and he is now paying the price for this break-up in world-wide disorientation and dissociation.[10]

This, then, was the framework of therapy and even salvation into which the Depth Boys stepped. The consumer's very dissociation of sensibility provided the means, and the justification, for 'organising chaos' as Bernays had put it. Like his predecessor, Dichter pushed the idea of the augmented product, but with even more psychological oomph. He had a penchant for presenting sophisticated ideas with disarming charm, for example: 'To ladies, don't sell shoes. Sell them sexy feet.' He was a master at motivational manipulation. In his 1947 book *The Psychology of Everyday Living*, he describes how soap is basically just lanolin, but it can be sold by 'extensive advertising campaigns to convince the male of the species that he suffers from body odour, and that this is why most girls reject him'. Similarly, women are shown 'moving picture stars, who stress the fact that they acquired their beauty primarily by using Lux toilet soap'. With brazen candour Dichter adds: 'The psychologist is not interested in the truth or falsehood of such claims. He is interested in the fact that a simple everyday commodity like soap has been advertised as a means of satisfying the important human need for beauty and happiness.' The circularity of the argument seems to have escaped him. After all, was it not the psychologist in the first place who created the need by stirring up insecurity? The name of the marketing game, then, was to 'trigger' buying behaviour. The advertiser baits the emotional hooks – freedom, love, sex, fear, envy, greed. He drops them deep in the psyche and waits for the little fish to bite.

Both Eddie Bernays and Ernest Dichter worked extensively with the automotive industries, big oil, and tobacco. Bernays first demonstrated the marketing power of the publicity stunt during the Easter Parade through New York in March 1929. Led by a lady called Bertha Hunt, a group of women who appeared to be suffragettes passed the Baptist church attended by John D. Rockefeller and each pulled from their suspenders a Lucky Strike cigarette. In front of flashing press cameras (that just happened to be placed there), they lit up, telling reporters that in so doing they were lighting up 'Torches of Freedom'. They said they wanted to 'smash the discriminatory taboo on cigarettes for women' and pledged that their gender would 'go on breaking down all discriminations'.

What the public didn't know was that Bertha Hunt was Bernays' secretary. He had set the press up for a photo opportunity complete with product placement. His stunt made smoking look 'cool' for women and linked Lucky Strike to that all-American virtue, freedom. Later, Ernest Dichter would follow on in similar vein. 'Smoking is fun,' Dichter wrote, in presenting his research for the tobacco industry. 'Smoking is a reward . . . I blow my troubles away . . . With a cigarette I am not alone.'[11] Here smoking becomes an antidote to anxiety, to loneliness; an emotional anaesthetic.

The Hidden Persuaders cites many similar distasteful case studies culled from an era before the advertising industry did a PR job to clean up its own image. Packard quotes Dichter from 1951 blatantly saying that a successful ad agency 'manipulates human motivations and desires and develops a need for goods with which the public has at one time been unfamiliar – perhaps even undesirous of purchasing'. Other ad men of the same vintage waxed lyrical on how 'the automobile tells us who we are and what we think we want to be'. The choice of gasoline 'helps a buyer to answer the question "Who am I?"' And generally, 'You want the customer to fall in love with your product and have a profound brand loyalty when actually content may be very similar to hundreds of competing brands.'

One of Packard's most disturbing case studies describes a market research project presented at a University of Michigan advertising conference in May 1954. The ad agency's remit had been to find out why little girls didn't get their hair permed. It turned out that mothers considered perming to be sexually precocious and some feared that the chemicals might damage their daughters' hair. With the help of child psychologists, the agency devised a series of projective tests. These explored children's responses to images of little girls standing at a window. The findings purportedly revealed that straight hair was associated with being lonely, curly hair with popularity. This resulted in a TV commercial aimed at opening up the perm market to children. It did this by persuading mothers to identify with their daughter's burning questions: namely, 'will I be beautiful or ugly, loved or unloved?' The answer, of course, was to be found in curly hair and therefore, in the profits of the chemical companies.[12]

* * *

In 1980, when I studied marketing as part of an otherwise excellent MBA course at Edinburgh University, none of this was on the core curriculum. The world guru was and remains Philip Kotler, Distinguished Professor of International Marketing at the Kellogg School of Management. To this day his bland tomes are the staple diet of most management schools. Back in the 1980s Kotler at least made passing reference to Dichter. Today, in almost a thousand pages of *Principles of Marketing* co-published with the *Financial Times,* there's not a glimpse, still less of Eddie Bernays or the use of an expression like 'motivational manipulation'. There's only squeaky clean descriptions of the 4 Ps – product, price, place and promotion – like eating a dry Weetabix with the box still round it. My suspicion is that the industry no longer wants to embarrass itself by teaching the revealing stuff. In any case, the industry already embodies the values that we have been looking at, but unconsciously. They arise spontaneously from the culture that marketing has normalised. Let me give an example and, in so doing, suggest just how far and to what depths colonisation of the soul can go.

Some years ago I carried out research into the advertising of Gallaher's two main cigarette brands in Britain – Benson & Hedges and Silk Cut. The findings were published as an occasional paper of the Centre for Human Ecology in 1996. They were picked up on by the BBC, *The Sunday Times*, *New Statesman* and *The Wall Street Journal* – with the latter featuring the paper as part of a front-page lead story.

My attention was first grabbed by a series of surreal ads for B&H that seemed to defy all the norms of advertising. These had picked up an array of industry awards. No fewer than five of them had been chosen by three dozen industry experts for inclusion in a book, *The World's 100 Best Posters*. These showed:

- The pack sitting like a trap outside a mouse hole.
- The pack perched like a bird inside a cage.
- The pack underwater, opened like a tin of sardines.
- The pack forming one of the great pyramids of Egypt.
- The pack against a torrid sunset, being carried away by ants.

The only printed message on each ad was the government warning: 'Cigarettes can seriously damage your health.' Equally bizarre were ads for Gallaher's Silk Cut range. These became surreal from 1983, starting with a poster that simply showed the health warning and a length of purple silk with a slash in it – silk cut. A relentless campaign that followed on for nearly two decades used a stream of images, typical of which were:

- A human figure showering behind a purple silk curtain.
- Surgical scissors with a wisp of purple silk lined up against a prison wall.
- A Venus flytrap plant monstrously ripping the crotch from purple silk trousers
- Four purple silk figurines shaped like stubbed-out cigarettes with chess pawns for their heads, lined up outside a toilet door that has a knife hanging on it.

- During the 1996 Edinburgh Festival that Silk Cut were sponsoring, purple bagpipe-like haggises, wandering in a bleak landscape of open-jawed mantraps.

Again, the only only indication of the nature of the product being advertised was the government health warnings. What could be going on here?

What caused my paper to make a media splash is that I suggested that sadomasochistic psychology was being exploited. The B&H sequence was made by the advertising agency Collett Dickenson Pearce. It was almost as if CDP were saying: 'Smoke these, and be. . .

- *trapped* like a mouse,
- *caged* like a bird,
- *canned* like sardines,
- *mummified* like the Pharaohs, and,
- *consumed* as carrion by the ants.'

Silk Cut's images from Saatchi and Saatchi seemed even more morbid. Masterminded by Charles Saatchi, the marketing man credited with bringing Margaret Thatcher to power, the campaign was unceremoniously dubbed 'Silk Cunt' by industry wags. But it seemed to be about more than just sex. There was an undertone suggestive of rape; even, like with the B&H ads, of death. Yet the adverts for both brands were outstandingly successful. A 1996 study by the Cancer Research Campaign showed that both brands had higher recall than any of its competitors amongst girls. Silk Cut topped the league with girls who had never smoked before. The ads may have been bizarre, but something about them was compelling too.

It's always fun in this kind of research to speak to the kind of people who are paid to tell you nothing. I therefore contacted Gallaher and spoke to Colin Stockall, their Media Services Manager. 'What's going on here?' I asked, not flinching from posing a leading question in the short opportunity during which time I hoped to hold his attention before he rumbled me. 'What's the meaning of, for example, the purple curtain? Is it an allusion to Hitchcock's *Psycho* where she's stabbed in the shower?'

'Well,' he said in a heard-it-all-before tone, 'I know some people interpret it that way but I can't say that's our view of it.'

It wouldn't be, and he wouldn't be able to say it was, so I chanced my luck elsewhere. This time I got on to the creative executive at M&C Saatchi – a spin-off company of the Saatchi brothers which had devised the haggis advert. 'What's the story with these extraordinary haggii?' I asked. 'Might this be how Gallaher sees its customers?'

He, too, had heard it all before but was less guarded. 'I think people would have to be either very negative in their view of life or overanalysing it to create a sub-plot that doesn't really exist,' he told me,

> I mean, the idea really is that these are not people, these are not living breathing animals. They're just objects that look funny. That look although they almost want to get trapped because that's what man traps do. They trap things. And that's what animals do. They step in things. You know, especially like dumb sheep-type animals. But these are more than that. They're just odd-looking, bagpipes, which have been made to look like haggises. It's a fantasy. It's just an odd image, and because it's odd it looks interesting. It captures people's attention.

Duly reprimanded for my negative view of life and yet fascinated by the confused tension in his account, I asked about the shower curtain image. 'Oh yes,' he acknowledged, '. . . people recognise the connection between the advertisement and *Psycho*, the thriller, so people think they're quite clever. It's smart arse. It affirms their intelligence and their wittiness. It strikes a chord with them.'

'And the pawns outside the door with the knife?' I ventured. 'It looks to me like they're dying for the loo. . .'

By now he sounded remote, lackadaisical, ready to get back to whatever he'd been working on. 'We call it *Dying for a Slash*,' he said.

* * *

My penetration of the bastions of CDP was also lackadaisical, indeed, serendipitous. Some of the finest anti-war advertising copy of the First Gulf War had been run by Amnesty International and Friends of the Earth. These full page newspaper spreads had been drafted by an Anglo-Indian advertising creative, Indra Sinha. We met in the course of campaigning against the war over the internet and he invited me to visit when next in London. Imagine the surprise as I went into a plush office block to see all Indra's right-on stuff there on the wall and, alongside it, ads promoting the British Army and B&H!

'I'm sorry, Alastair,' he said with a knowing shrug, 'but something has to pay the bills.' Indra was only one of a much larger team. His was just one voice in the composition of some of the ads. But he told me how they'd hit on the idea of using surrealist images after the government had tightened the rules making it impossible to show cigarettes in much of a positive light. Apparently one of the CDP chiefs described the campaign as 'very Jungian', but nobody imagined that the images were doing anything more than just being 'clever'.

I replied that 'clever' doesn't just come from nowhere. In 1924 the French psychiatrist and cultural critic André Breton had defined surrealism as 'pure psychic automatism by which one proposes to express . . . the actual functioning of the mind'. It is 'the future resolution of . . . dream and reality . . . into a kind of absolute reality, a *surreality*'.[13] The resonance with the slightly earlier concept of Caledonian antisyzygy will be clear. My question was about the 'actual functioning' of these macabre images. I told Indra what my theory was. In the final stages of Sigmund Freud's thought in the early 1920s he arrived at a point of view that embarrassed even many of his supporters. His medical work treating the battle neuroses of traumatised soldiers led him to conclude that, in addition to *Eros* – the life force – there was also its counterpoint expressed as a 'death instinct'. This was later called *Thanatos* after the Greek god of death. 'In the end we came to recognize sadism as its representative,' Freud wrote, 'and life itself would be a conflict and compromise between these two trends.'[14]

Was it possible, I wondered, that at some level in Gallaher's creative hierarchy, minds had wandered where no advertiser would have previously thought or dared to venture? Was it possible that the advertising industry, having worked all the more conventional motivating forces to death, had hit on death wish as the final frontier? The Silk Cut ads make it even plainer than those from the B&H stable. With Silk Cut, death, sex and the scatological unambiguously interweave. I wondered if such surrealism could be capitalising on the nihilistic roots of addiction in the gnawing emptiness of the hollowed-out soul? After all, there is a well-recognised mystic link between sex and death. Mythology is full of it. The French call orgasm '*la petite mort*'– the little death. There can be a sense of release in the prospect of death. Some analysts of women's erotica suggest that rape fantasies can provide an excuse to liberate an inhibited imagination. Could Gallaher, wittingly or unwittingly, be turning the government's health warnings into a unique selling point? Could the adverts be saying, in effect, that smoking offers everything they've ever warned you against? Cigarettes kill. And if you want *la petite mort* big time, then why not just die into the ecstasy of Gallaher's arms? As the Roberta Flack lyrics put it, 'Strumming my pain with his fingers | Singing my life with his words | Killing me softly with his song | Killing me softly, with his song.'

Indra was fascinated. He said that they never discussed the ads in terms of 'meaning'. The guys who design these things are creatives who come up with images, not intellectual propositions. But now that I mentioned it he could glimpse something that was possibly there. He said that maybe the very reason why he and his colleagues were good at their jobs is that they don't think too much. They just get given a brief and follow their gut feelings.

Indra was perhaps more open than many would have been to what I was suggesting. One of his other claims to fame is having translated the *Kama Sutra* from the original Sanskrit. In the Tantric traditions of esoteric Hinduism, the connection between sexual love and death is celebrated. One is seen as being the flip side of the other. Both are equally sacred. The initiate learns to see through death and into the love beyond. In a similar vein, but without overt sexualisation, the great mystics of the world talk of 'dying into love'. Mystical death is therefore the essence of advanced spiritual experience. The outward form of this varies between traditions, but inwardly it's the meaning of the Cross. As the Spanish mystic St John of the Cross said in one of his poems: 'In hope I now begin to die | My destiny I seek, for I | Am dying, so as not to die.'[15]

And that last line is the essence of it. Mystical death is not about running into the buffers of terminal morbidity. Mystical death is the small self of the ego letting go into the great Self that is grounded in the divine. It is the soul's marriage to God, like we see so exquisitely expressed in the biblical *Song of Solomon*. Such is the apogee of 'liminal' experience – the crossing of a threshold of consciousness into higher reality. But consumerism, the wares of tobacco companies and other forms of addiction, don't offer real liminal experience. They just offer a transient glimpse. They merely offer the 'liminoid', which is a false semblance of the liminal. As the Franciscan spiritual teacher Richard Rohr describes the distinction: 'The liminoid is a movement into trance and unconsciousness so nothing real will be revealed . . . The liminoid feels like the real thing, it feels momentarily renewing, but it is just a diversion and actually reaffirms our ego . . . and our capacity for denial. It is not a threshold at all, only more of the same.'[16]

Could this be the core dynamic by which consumerism sustains itself? Addictions are powerful precisely because they taunt us with our heart's longing. But they fake it. They short-circuit and actually block off the real thing – the focus of our ultimate concern. The great American mythologist Joseph Campbell, whose ideas helped to structure movies like *Star Wars*, said that there is only one basic form to all the world's great stories. It's tripartite. The hero goes on a journey of departure, initiation and return. That 'hero' is every one of us. When we first set out on life's journey, most of who we are is shaped by where we've come from – the inheritance of our family, education and community. We're in our small selves and hardly yet know about the great Self. Then we hit the trials and tribulations of life. It might be childbirth, bereavement, broken love affairs, a devastating fight or several of these things one after another. What matters is not the outcome. As Kipling said in his poem, 'If', such outward semblances as triumph and disaster are 'impostors just the same'. There is only one thing that really matters and that is to find the heart's courage. The very word 'courage' comes from the French *cour*, meaning 'the heart'. It is the courage to endure. The courage to trust to life's deep processes. The courage, always, to love. Faith, hope and charity.

Only then will the small self of the ego find its safe and rightful harbour in the ocean of the soul. Only then can we be ourselves and enjoy ourselves with hubris quelled. Only from such passing through the fire that burns off the superfluous and refines the essence are we equipped to return as elders to our communities. And this matters in the healing of the world. It matters from that level of the psyche at which all things interconnect. It gives meaning even to the little lives, to the sick, the wounded, the disabled and the bedridden. It matters even, and perhaps, especially, in the ordinary heroics of those who never get noticed and are always passed over by the calculating ways of an opportunistic world. For something new is born in each and every one who faces a rite of passage and says to life: 'Let it be.'[17] Let it happen. Let it be me, here, now that opens my empty womb to be infused by the Spirit. As Joseph Campbell sums it up, 'The effect of the successful adventure of the hero is the unlocking and release again of the flow of life into the body of the world.'[18] Such incarnation is the human condition's solitary hope.

Here too lies exposed what consumerism blocks us from. It interrupts the very journey of life. It keeps us narcissistically at a child-like level of immaturity, seeking only the next fix. We bypass challenges that might heal our emptiness and give birth to human potential. My friend Indra is exceptional. He phoned me up a year later and said, 'I just wanted you to know I've left CDP. I've decided to stop being a pedlar of death.' He bit the bridle of poverty and became a novelist instead. One of his books, *Animal's People*, is described as 'a stunningly humane work of storytelling [that] challenges us to define what it is to be human.' Imagine my delight when I saw it shortlisted for the 2007 Booker Prize! But other opinion leaders in our culture perhaps have more to carry on their consciences. Charles Saatchi held on to Silk Cut long after he had dropped most other accounts. With his ceaseless buying and selling he's been credited almost single-handedly with defining Britart – the Young British Artists movement. He describes such art as 'neurotic realist' and himself admits to being a 'gorger of the briefly new . . . ghastly, but true.'[19] When *The Guardian* asked if he had happy memories from the halcyon days when his gallery stood on London's South Bank he said:

> My best memories of County Hall are that we were very popular with schools. We must have had 1,200 to 1,300 schools come. I know I sound like some ghastly creep, but there is something enchanting about seeing groups of children sitting round a Chapman brothers piece with penises coming out of girls' eyes, drawing it very neatly to take back to their teachers.[20]

And there I rest my case. We will probably never know what went on at senior levels in Gallaher and its advertising agencies. I certainly don't imagine that many ad men today hang around wine bars reading Freud. But could Charles Saatchi have done so? Could the very unspeakability of his ideas have been the secret of his success? We can conclude but one thing with certainty. The same mind in charge of 'Psycho' and 'Dying for a Slash' also messed with schoolgirls' minds. Lethally so, given the Cancer Research Campaign findings.

Let me be clear that my argument here is not with those who smoke cigarettes. It is with the pushers. If one Googles 'Silk Cut' my paper on 'Eros and Thanatos' in Gallaher's ads comes up, complete with a gallery of the images we have been discussing, immediately alongside the company's own website. They must be right pleased! I also noticed that, after the blaze of publicity surrounding the paper's publication in August 1996 through until February 2003, when cigarette advertising was finally banned in Britain, most of Gallaher's ads were fairly muted. But one of them did bring a wry smile to my face.

It was February 1997. There, on a huge billboard by a football stadium, I saw the familiar cigarette pack. It had been cut in half and sewn back together again with big purple silk stitches as if it had gone through a surgical procedure. I admit that my imagination ran a bit wild. For a moment it felt as if old Charles was conceding something; as if he was saying, 'stitched up'!

* * *

I have tried to suggest in this chapter that there may be depths to what it means to be a human being of which most of us are only dimly aware. If we persist in choosing ignorance we do so at our peril. I am, of course, far from alone in making this observation. We have seen what Carl Jung and John Kenneth Galbraith said about it several decades ago. Erich Fromm also summed it up in his question, 'To Have or to Be?' But perhaps the most unexpected voice of concern came in President Eisenhower's farewell address to the nation of January 1961. This former General of the Army warned that something 'new in the American experience' had emerged out of the twentieth century. He named it the 'military-industrial complex' and said that it sought 'total influence – economic, political, even spiritual'. That much is well known and oft quoted. Less well known is that he extended his analysis to the spiritual and political threat of our own consumerism. He said:

> As we peer into society's future, we – you and I, and our government – must avoid the impulse to live only for today, plundering, for our own ease and convenience, the precious resources of tomorrow. We cannot mortgage the material assets of our grandchildren without risking the loss also of their political and spiritual heritage. We want democracy to survive for all generations to come, not to become the insolvent phantom of tomorrow.[21]

Today's climate change could develop into stunning vindication of those words. The consumerism that Eisenhower pointed towards comes from vulnerabilities opened up because our inner lives have been collectively drained. That's what allows the liminoid to be passed off as the liminal. It perverts all that is most sacred in our humanity. That's the meaning of 'vice'. It grips and squeezes out the lifeblood.

I have chosen tobacco here as a case study. But examples also abound with the alcohol, food, fashion, travel and automotive industries. From an environmental point of view the health warning on tobacco products could be more widely replicated. Here's my suggestion for the wording:

This product can seriously damage
your planet causing loss of life
and species extinction

And that is where our study of hubris has led us to. The roots of climate change touch so many aspects of our lives. We are all complicit to varying degrees. The challenge is to humanity as a whole and not just to the most blatantly profligate or manipulative. The question of how to tackle global warming is about technical, economic and political considerations for sure, but more fundamentally, it is about how the soul can be called back; the modern soul, like that of a latter day Gilgamesh wandering restlessly amongst the hungry ghosts of insatiability. That is what I will now attempt to address.

Chapter Eight

JOURNEY INTO THE SOUL

We have seen that consumerism – which is to say, consumption beyond the level of dignified sufficiency – has not come from nowhere. Its cutting edge is hubris. Its ricochets of violence reverberate around the soul. There is nothing new in such an observation. 'You were in Eden, the garden of God,' said the Biblical prophet Ezekiel, 'but in the abundance of your trade you were filled with violence, and you sinned.'[1] The only difference these days is scale. Consumerism has become to the environment what the H-bomb became to war. Both of them challenge us all to think outside the box of bygone behavioural norms.

Violence is blind, senseless and numbing. We can see that clearly enough when red blood flows, but otherwise we frequently lack appreciation of what violence actually looks like. It's hard to see its subtle manifestations, yet they're all around. To invest money in ways that exploit the desperation of others is a form of violence. To consume what harms the Earth's life support structures is a form of violence. To substitute the liminal with the liminoid is the deepest structural violence. This is what the prophets call idolatry because it sets up false 'gods' – false hopes for meaning and purpose in life. Yet it's often hard to see all this. Violence blames its own victims: it reproaches even the blindness of those whose eyes it has put out. We don't want to see it and sometimes we just can't, because violence is such a powerful agent of consciousness change – a form of hypnotic trance. Under its spell we 'go under'. As we discussed earlier, it's one of the factors that keep us constrained within 'consensus trance reality'.

In *Soil and Soul* I discussed how hypnosis has been defined as 'a contraction of the usual frame of reference'. When our awareness closes down empathy also contracts. Wittingly or unwittingly we therefore become open to violence and thus the irony of how readily the oppressed becomes oppressor. As we have seen, I consider violence to have been a major factor underlying Eliot's 'dissociation of sensibility' where the 'inner' and the 'outer' aspects of our lives move out of synch. If just the 'outer' aspects of need are addressed, then the 'inner' component must remain a hungry ghost, forever unsettled and wanting. We may shop until we drop, but as Mick Jagger said, we 'can't get no satisfaction'. We can champion ideas like 'green consumption' and where these promote ways of living within nature's carrying capacity, they are most certainly central to the solution. But where the consumption in question is a mask for mere consumerism – where it hides its real motive behind the lie of a feel-good veneer – it is but greenwash and merely compounds environmental problems with the fog of deception.

Let me give an example. I have before me the autumn 2007 edition of a consumer magazine called *The Quarterly*. It carries a double page advertising spread for Land Rover 4x4s. The caption boldly reads: 'GO RESPONSIBLY: All new Land Rovers are fully CO_2 offset for 45,000 miles.' Well, I've checked on their website and frankly I don't believe that this is true in any way that most ecologists would consider credible. The advert gives no explanation of the supposed carbon offset. The small print merely states that a Range Rover's urban fuel economy (prices starting from £55,000) varies from an appalling 12.6 to an equally unacceptable 19.6 miles per gallon. (An efficient modern car achieves three times that.)

What we see here is only one of the more overt examples of advertising hypnosis. Dangled before the prospective customer is the promise of absolution from eco-sin. Like in the medieval church, an eco-indulgence is offered up for sale: a small portion of a vehicle's price goes to clean energy projects in the Third World. What gets offset is not so much the car's massive carbon footprint, but any lingering ethical qualms in the consumer. As such, the consumer's frame of reference is refocused and in a way that contracts it from the big picture. The customer pays attention to the pretty assistant while the magician, a master of greed-induced trance, pulls his sleight of hand. And hey presto . . . out from the forecourt pops yet another satisfied punter. And not only that, but for those with eyes to see the buyer's insecure lack of confidence is paraded on high. I'm not making that up. If we care to unpack its own words, the Range Rover website tells us so! It boasts that the 'acclaimed Command driving position elevates the driver above traffic for a sense of confidence and ease that is without equal.'[2]

Six months later the company seemed to have toned down its claims in the glossy magazine ads. The 'offset' statement was still there but minus any suggestion that it is being undertaken 'fully'. This time a Freelander 2 is shown ripping across open countryside following a girl in a hot air balloon as its owner boasts about the need for 'a pretty good chase car – one that can handle dirt tracks and streams . . . down in the back of beyond'. And the main caption proclaims: 'I've always liked to go places where there aren't any roads.' So much for the shelf life of 'GO RESPONSIBLY'!

Personally, I don't buy any of it – neither the cars nor the way that they are sold. A company pandering to this sort of hypocritical psychology may have succeeded in cutting its manufacturing carbon emissions by 30% as some of Land Rover's ads claim to have done, but they cannot yet lay full claim to corporate social responsibility. Their pushing of a luxury product and the way in which that raises the aspirations even of those who cannot attain what is being advertised pollutes not just the atmosphere, but our minds. This is the kind of corporate marketing behaviour that we must name and shame. The prestige normally associated with such chattels needs to be inverted. Apart from people like farmers who really do need to drive off-road, over-engineered vehicles need to be seen for what they really are – a source of embarrassment.

Recently I witnessed a situation in which a powerful man actually was embarrassed by his own car. He was the former leader of a major British political party and we were both part of a discussion panel. On the way up to the podium he had cause to explain why his wife happened to be in the huff. 'We've come up in the little car and she's not pleased,' he said, in exasperation. 'But I told her there's no way I was turning up at an environmental conference in the Jag!' Here we see an example of the negative drive of 'push'. He would rather not have had to leave the Jaguar behind when going on a long journey, but implicit social pressure had pushed him so to do. That is one part of the shift in social attitude that needs to come about. But negative drives alone are not enough. They carry a counterproductive shadow. Another time somebody of this man's standing could simply decline coming to a 'green' gathering. That would be a loss because it would diminish the movement's capacity to connect with the mainstream. It would reinforce the division of the world into, if I might generalise wickedly, the Jaguar set and yurt-dwelling yoghurt knitters.

In contrast, the positive drive of 'pull' comes about when people see that they can be truly enriched from being drawn into a new mind and a new heart. This is waking up – the antidote to consensus trance reality's hypnotic daze. It requires opening to new possibilities in life that are best summed up with the word 'presence' – the awareness that makes connection and empathy possible. Here lies transformation from the false liminoid to the true liminal; the inauthentic to the real. It is spirituality practised in a very grounded way that re-endows our world with attention.

What's important to remember is that consensus trance reality is there for a reason. Most of us can't be conscious of everything all of the time. Our attention is limited, therefore the 'hypnotic' capacity for it to focus or be focussed and close down to wider reality is very necessary. Psychologists call this the 'automatisation' of perceptual and cognitive structures. It's like when you learn to ride a bike. At first you can think of nothing else but how do you manage to stay on! But as you get good, you 'forget' what you're doing and it all comes automatically. This frees one's capacities for other tasks – focussing on the cycle race, enjoying the flowers or chatting to a companion. The problem is that this useful automatisation can go too far. It can block us from seeing the big picture. Without realising it, automatisation can cause us to become frozen in perceptual and cognitive patterns that sustain dysfunctional behaviour. And as we've seen, manipulative forces can hijack the process and use it to colonise the soul.

All this makes it important, at least now and again, to 'de-automatise' – to blink, wake up, and trip out of the trance. As William Blake put it in *The Marriage of Heaven and Hell*, 'If the doors of perception were cleansed, every thing would appear to man as it is, infinite. | For man has closed himself up, till he sees all things thro' narrow chinks of his cavern.'

Arthur Deikman, who is professor of psychiatry at the University of California, sees deautomatisation as the essence of mystic or mystical experience.[3] It means re-endowing the everyday world with attention such as lies at the heart of much spiritual practice. It requires 'mindfulness' in everyday things – even just walking along the street or washing dishes – but doing so with attention. Personally I find it incredibly difficult to achieve for any length of time, but it certainly brings out the difference between absence and presence to immediate reality. The reward is that it helps us to reorder priorities and brighten life. For example, it can enable a shift in attitude from seeing the kneading of bread as a chore to being the pleasure of working one's muscles, and even the kneading-in of love. As such, homemade bread can have an augmented value that is very different from that usually bought in the shops. It can touch on the sacramental, and this added value comes for free.

The old notion of 'saying' grace before a meal is another example of presence. Such a ritual, which doesn't have to be 'said' as such, connects the practicalities of the meal – that which pertains to the *Logos* or the 'head', with the *Mythos* that embeds it in story – the role of people, life processes, and even the cosmos in providing sustenance. It integrates gratitude – which, rather surprisingly, is where the qualities of *Eros* come into play. True *Eros* is about much more than sexuality alone. True *Eros* is the embodied experience of the engaged heart. It is the mediation by which the inner and the outer, the spiritual and the material are united. As the black feminist philosopher Audre Lorde puts it in a vitally important essay, *Uses of the Erotic*: 'The dichotomy between the spiritual and the political is also false, resulting from an incomplete attention to our erotic knowledge. For the bridge which connects them is formed by the erotic – the sensual – those physical, emotional and psychic expressions of what is deepest and strongest and richest within each of us, being shared: the passions of love, in its deepest meanings.'[4]

It is this erotic capacity to have an embodied relationship with reality that gets lost in the dissociation of sensibility. Indeed, T.S. Eliot's essay spells out how the older metaphysical poets had the 'essential quality of transmuting ideas into sensations, of transforming an observation into a state of mind'. He adds that in seeking to recover sensibility one must seek in the heart, for sure, but also 'one must look into the cerebral cortex, the nervous system, and the digestive tracts.'[5] In other words, the reconnection with life that we need if we are to rise from consumerism to grace is a practice that engages the entire mind-body system; the totality of the psyche in nothing less than its cosmic grounding.

Violence can never achieve this because it violates psychosomatic integrity and, thereby, breaks the heart. This is why, as theologian Walter Wink shows in his study of power that has informed so much of my own activism, the idea of 'redemptive violence' is always a myth.[6] Violence ruptures the deep listening and receiving that comprises empathy. In *Eros* is its healing, and so the 'passion' of Christ (or of Socrates) was to face suffering with a greatness of soul that sidestepped further perpetuation of the spiral of violence.[7]

The true erotic as Audre Lorde describes it is therefore the capacity to experience 'sensation' in a manner that is fully integrated – 'with the heart engaged'. In contrast, she shows that the merely pornographic 'is sensation devoid of the heart's engagement'. In this analysis we can see that Lorde takes us far beyond the normal sexualised understanding of the pornographic. She opens up a generalised insight with which we can unmask the core dynamic of consumerism. Consumerism is in the broadest sense pornographic precisely because it promotes sensation *without* engagement of the heart. In contrast, consumption can be erotic, all the way to grace-filled sacrament, wherever the body's sensation and the heart's empathy are brought into harmony together. Such presence is a grounded here-and-now spirituality. It allows us to enjoy *Logos*, *Mythos* and *Eros* to the full and in all walks of life.

The antidote to being entranced by consumerism is, therefore, to face up to our own complicity in pornographic living and to awaken to the alternative. By gazing at Truth in this way we can become open to the possibility of moving, and being moved, beyond our liminoid addictions. The world is re-enchanted by this affirmation, release and relishing of *Eros*. It points towards a transformation of the entire mindless economic system and its replacement with what E.F. Schumacher in *Small is Beautiful* called the 'Buddhist economics' of mindfulness. Here the inner vision of the heart synchronises with the outer resources of the planet. In very practical terms it means, when we are able so to do, choosing the products of such efforts as Fair Trade and organic agriculture, things that are locally made, mutual and cooperative business models, community land ownership, education for creativity, ecological architecture and so on. It's simply about aligning our consumption to a path with heart. In so doing we need to try and dissociate ourselves from the temptation, often unconscious, of aping the mores of the rich. The Earth can no longer afford the rich. All of us must turn from being pedlars of death to seekers of life.

But there is a strange twist in trying to achieve this. To choose life we have to get more real about what is, on the surface of things, its opposite: namely death. We have to move beyond Gilgamesh's frenzied denial of mortality and, instead, find the courage to face death in forms both literal and metaphoric. Only then can we unreservedly embrace the deeper strata of life. Before this chapter is through I hope to demonstrate more of what I mean. But first, some practicalities about planetary death.

* * *

WWF point out that for everybody in the world to live like an average Indian would require half a planet. For us all to live like Europeans would take three planets. And like Americans, seven. Such planetary impact faces us with what, in the past, has been called a 'dying time'. As we saw earlier, over the next couple of centuries climate change has the potential to cause death on a massive scale to both human beings and entire species. The 1,800 people left dead and another 700 missing after Hurricane Katrina struck New Orleans and Louisiana in 2005 could be just a trailer for more to come. As for our planetary neighbours, the International Union for the Conservation of Nature – the world's most prestigious nature conservation body – maintains a Red List of Threatened Species that, in 2007, listed 16,306 plants and animals threatened with extinction.[8] More get added every year as a wide range of pressures cause further habitats to be destroyed. Climate change is only one of these pressures, but as we have seen, the more that temperatures go up the greater the extinction risk. The models used by the IPCC suggest that a doubling of CO_2 levels would produce a temperature rise of about 3°C.[9] Depending on what remedial measures world leaders might put in place, this and probably an even greater temperature rise is likely to come about during the present century. The question is where will it all stop, especially if tipping points start adding huge volumes of methane to the equation?

Once again, the horrible irony is that some of those parts of the world most responsible for climate change are the least likely to suffer in the short to medium term from the direct effects. Today's global warming may be dramatic in terms of past geological precedents, but even a centimetre's rise in sea level every three years, as is happening at present, is glacially slow in terms of human politics. But we cannot presume that we who can perhaps ride out the storms awhile will always stay immune. Wider factors come into play that could take us all by surprise. When any animal exceeds its environmental carrying capacity, nature sorts it out in one of three ways – war (or predation), famine or pestilence.

Pestilence, which is to say, disease, usually has the most catastrophic impact on populations over wide areas. If we examine a graph of world population levels throughout human history we see that very little changed until the late Bronze Age when agriculture got well underway. Thereafter, the number of people on the planet rose gradually, but it was not until early modernity and the advent of industrialisation that the numbers took off exponentially. During the reign of King James, worldwide population stood at only about half a billion – 500 million. Even as late as 1800 it had reached only 1 billion.[10] By 1955, when I was born, it had risen to 2.5 billion, and in July 2007 it reached 6.6 billion.

Throughout the past millennium there was only one major dent in this relentless rise. That was the Black Death around 1347. Europe lost one in three of its people and the impact on other parts of the world can only be guessed at. In the previous millennium, the Justinian Plague that started in 541 may have originated from privation following the extreme cold weather of 535–36. This was most likely caused by a 'volcanic winter' where dust dimmed sunlight after a massive volcanic explosion – possibly the one that created the Rabaul crater in Papua New Guinea.[11] The resulting famines and plague are mentioned in a variety of ancient manuscripts. For example, the Irish *Annals of the Four Masters* states that in 543 'there was an extraordinary universal plague through the world, which swept away the noblest third part of the human race'.[12]

More recently, the 'Spanish Flu' that broke out at the end of the First World War in 1918 produced a global mortality usually given as 20 million, but possibly as high as 40 million.[13] The war killed 15 million in four years, but the flu, with young men and pregnant women its chief target, brought down double that in six months. It even spread to remote Pacific islands and, while mortality amongst the infected was generally 2–20%, amongst the Inuit, who had no resistance, mortality was 100% in some communities. In total, up to 5% of the human population may have been wiped out, mostly on contracting pneumonia as a secondary infection.

Crudely expressed, human impact on the planet is the product of population and consumption. The rate of growth in world population is slowing and is expected to stabilise around the middle of this century at around 10 billion. But with rates of consumption escalating and climate change adding to the pressures, life support systems are likely to become more stressed. This begs the disturbing question as to whether pestilence might be a factor that could take our need to reduce carbon emissions in hand for us. Imagine, for instance, that the HIV virus causing AIDS had been spread not by sex, which can be a relatively rare activity, but by sneezing! Had that been the case, then by now readers of this book would be numbered amongst the very few remaining survivors of humankind. Similarly, as of January 2008 the H5N1 variety of bird flu has infected only 353 people worldwide since 2003. But it has killed 221 of them – a staggering 63% mortality rate.[14] Type A influenza viruses such as this mutate very easily and quite without warning. Were H5N1 to change from being a sluggish to a highly contagious strain in human hosts and yet maintain its present lethality, the effect would be devastating. As the world-renowned virologist Robert Webster put it in the title of an article published in *American Scientist* in 2003, 'The world is teetering on the edge of a pandemic that could kill a large fraction of the human population.'[15]

The more that ecosystems come under stress the higher the likelihood of this kind of thing happening. This is why many flu epidemics start in the Far East. Pressure on the land there favours the production of pigs and poultry that can be intensively reared. When new strains of disease do break out, high human population density makes it easier for them to establish a pool of infection from which to spread.

As the Earth's carrying capacity gets more and more stretched by consumerism and climate change the epidemiological risks will be evident. The World Health Organisation (WHO) takes these concerns very seriously. It is poised, if necessary, to block airline traffic in and out of any country where a dangerous epidemic takes hold. In 2003 international restrictions aimed at stopping the spread of Severe Acute Respiratory Syndrome (SARS) cost the economies of East and South-east Asia an estimated $18 billion according to the Asian

Development Bank. That's about $2 million for each person who became infected by the disease.[16] Worldwide, WHO maintains an Epidemic and Pandemic Alert and Response (EPR) division which monitors possible pandemics such as this. Its website pulls no punches:

> During the twentieth century, influenza pandemics caused millions of deaths, social disruption and profound economic losses worldwide. Influenza experts agree that another pandemic is likely to happen but are unable to say when. The specific characteristics of a future pandemic virus cannot be predicted. Nobody knows how pathogenic a new virus would be, and which age groups it would affect . . . However, even in one of the more conservative scenarios, it has been calculated that the world will face up to 100 million outpatient visits, more than 25 million hospital admissions and several million deaths globally, within a very short period.[17]

In assessing the risks of climate change, humankind therefore needs to be alert to the possibility that the planet could one day give a feverish shudder. It might suddenly impose limits to human environmental impact such as we have so far declined voluntarily. The 'blizzard of the world' could cross thresholds microbial as well as meteorological. Facing up to this is one aspect of what I mean by facing up to death. To do so is disturbing, but the trouble with the alternative – keeping our heads in the sand – is that we never know what's likely to creep up and get us from behind!

As such, facing the possibility of death is a re-grounding in ecological reality. But it's also potentially more than that. I believe it can re-ground our humanity. We are already witnessing the extinction of species and suffering like that in Darfur or New Orleans. Such scenarios are likely to multiply. I believe that one of the competences we therefore need to develop today is to become what I would call 'planetary hospice workers'. I think that T.S. Eliot captured it in his profoundly mystical poem, *Four Quartets,* where he said, 'the whole earth is our hospital' in which:

The wounded surgeon plies the steel
That questions the distempered part;
Beneath the bleeding hand we feel
The sharp compassion of the healer's art
Resolving the enigma of the fever chart.

Our only health is the disease
If we obey the dying nurse
Whose constant care is not to please
But to remind of our, and Adam's curse,
And that, to be restored, our sickness must grow worse.[18]

Hospices that provide palliative care for the dying sometimes describe a process that terminally ill patients may pass through. It is called the 'Kübler–Ross grief cycle' – a five-stage sequence of denial, anger, bargaining and depression that follow one another before the peace of acceptance is finally reached.[19] I cannot help but feel that this echoes many people's response to climate change. Those who understand the sequence and are willing to serve however they can, without indulging in gloom and despair (however understandable they undoubtedly are) – these are the planetary hospice workers.

Either people like Martin Durkin are right, and climate change denial is justified, or humankind needs to shoulder the burden of awareness that comes with the knowledge that we are now in a great dying time of evolutionary history. The question of whether technology, politics and economic muscle can sort out the problem is the small question. The big question is about sorting out the human condition. It is the question of how we can deepen our humanity to cope with possible waves of war, famine, disease and refugees without such outer wounds festering to inner destruction.

The Buddhist ecologist Joanna Macy says that we need to face the depression that such thoughts rightly evoke and enter our despair. However, finding the courage to do this moves us through into empowerment. Macy therefore sees the suffering that humankind is confronted with as a call to spiritual growth. She describes the journey on which we have embarked as 'the great turning' – 'the essential adventure of our time: the shift from the industrial growth society to a life-sustaining civilization.'[20]

When we can see it like this the burden of awareness starts to shift a little. Our hope and trust must be that it transforms into a 'precious burden' – precious because it illuminates, as we saw Audre Lorde say, 'the passions of love, in its deepest meanings'.

None of this means that we stop striving for environmental sustainability. Spiritual growth does not mean curling up to die or perversely relishing the suffering – especially from a distance! Our call is to continue doing all the things we need to learn to do – the recycling, the energy efficiency, looking at what we eat, how we travel, and so on. Even if these are not enough to make much difference now, they may become invaluable patterns and examples for some post-apocalyptic future. We therefore need to develop a peculiar combination of being both level-headed and idealistic. We need to walk with what the theologian, Oswald Chambers called, 'the Discipline of Disillusionment'.[21]

Chambers was a Victorian Scot who said that 'most of the suffering in human life comes because we refuse to be disillusioned.' But the disillusion in question is merely an engagement with truth, for 'to be disillusioned means that for us there are no more false appearances in life.' On the surface of things this could be seen as a recipe for nihilistic despair. But that's why Chambers calls for it to be a 'discipline' – something that one follows, as would a disciple who is moved by a higher cause.

The discipline in question is to hold the things that disillusion us in a framework of life that allows compassion to enter in through the cracks. This is the work of grace. As Chambers said, it 'brings us to the place where we see men and women as they are, and yet there is no cynicism, we have no stinging, bitter things to say.' It is in such a melding of humility and realism that joy can unexpectedly break through.

* * *

Vérène and I went to bed early on Hogmanay 2006. We didn't

stay up to see 2007 in because she was seven months with child. The scans showed it to be a boy. We had delighted to watch his little heart beating, to see his bones knitting into shape. And that afternoon Vérène had been playing music to him. She felt that he responded to its different moods. Who knows what this child might become. It was a very happy time; a time of wide open possibilities.

We had thought a lot about whether or not it was right to bring a child into a world such as ours. It's not just all the extra consumption, but also the ethics of creating life in a dying time. Both our youths had seen optimistic times. For me, maturation into adult consciousness had been touched by all the hope that the sixties had brought to awareness. The Woodstock song reminded that we are stardust, and golden, and that we've got to make it back to the garden.

As late as 1990 hope was still unfolding. Mandela was free and apartheid was in the dock. The Berlin Wall had come down and the mass uprisings that demolished it were remarkably nonviolent. The atomic war once feared had never come about, and neither the nuclear winter that might have followed it. Instead, politicians were cutting the military's share of national expenditure and discussing how to spend the 'peace dividend'. In Ireland and Italy population rates were, to everybody's astonishment, falling below replacement levels. Around the world people were more healthy and living for longer than ever before. Even Mrs Thatcher turned 'green' before leaving office, albeit joking that she wanted her statue cast not from bronze but iron!

But today, the contours of that optimism have shifted. Trident submarines have been scheduled for replacement, and we have created military disasters in Afghanistan and Iraq. Iraq alone has so far notched up a 'true cost' of \$3,000,000,000,000 according to the Nobel prize-winning economist Joseph Stiglitz in *The Three Trillion Dollar War*. Meanwhile, torture has again become routinised and, in 2005, the British government even tried (but thankfully failed in the courts) to have evidence so extracted rendered acceptable in law. Little wonder that our public spaces now catch the karmic retribution of terrorism coming home to roost. On top of such militarism, world economic growth outstrips most efforts to mitigate climate change. We in the West can blame the East, but a quarter of China's carbon emissions come from its exports to the West.[22] We've merely hidden part of our carbon footprint by exporting it.

Hope and despair therefore oscillate in the mind. On the one hand, the climate scenario for other species and for our own children's children – especially those living in vulnerable parts of the world – appears deeply worrying. On the other hand, who knows for sure what might happen in the world? The models and the scenarios are all so complex. There could be protective feedback mechanisms of which current modelling is unaware that will kick in and save us. After all, humankind has been gripped by exaggerated doom and gloom scenarios many times before. They're great for galvanising the hapless with new-found meaning and boosting the egos of their prophets. Could it be, then, that some of the wilder climate change fears are just a modern version of those folk panics; what social anthropologists call millenarian cults? Could it be that for all the fuss and bother, we're merely faced with the cult of climate change? This time it's not little green men or the more paranoid parts of the Bible that herald the end. This time, in keeping with the spirit of the age, it's a secular apocalypse with prophets in white lab coats warning of our wicked ways!

I think that some of the fringes of the climate change debate do reflect cultic tendencies, but it's scarcely credible to argue that for the scientific mainstream. As we have seen relentlessly throughout this book, peer-reviewed science anticipates a deeply disturbing long-term scenario for the world. But our capacity for social and political action is caught in the 'frog paradox'. The apocryphal frog – I rather doubt that it would be true – sits in a pan of water on the stove, but at no time until too late is the discomfort enough to make it jump out. That's the worry about what faces us. That's why we need to tackle the scientific concerns square on and yet not allow ourselves the indulgence of succumbing to despair and giving up on life. Islam teaches that despair is a sin. It's a turning of one's back on spiritual hope, and whether we're Muslim or not we can learn from that. It's why we similarly need to distinguish between despair and the Discipline of Disillusionment. It's why we need to hold fast to the grace of counting our blessings, including the many that science and technology have given us. For, as the man with toothache said when asked if he thought modernity had yielded any real benefits: 'Dentistry!'

These, then, were just some of the welter of feelings that Vérène and I had running through our minds in 2006, and somehow we both felt that it was important to bring this child into being. I already have two children. I delight in watching their wisdom unfold as their adult lives progress. They have helped to keep me alive to the world. And tomorrow's world, more than ever before, is going to need young people who have been well loved for themselves. It will need people who have been honoured in their primal integrity. Our problems are caused less by too many children than by not enough who have been really wanted. If there's to be a future worth having the world needs people who have learned how to love . . .

. . . And I think back to my own childhood and our primary school in the village of Leurbost on the Isle of Lewis. I'm remembering that we had a music teacher who had a little white dog. He used to teach us the songs of Robert Burns . . .

O, my Luve is like a red, red rose,
That's newly sprung in June.
O My Luve is like the melodie,
That's sweetly play'd in tune.

Burns wrote *My Luve is Like a Red, Red Rose* in 1794. It's about romance, but here is a love that deepens into strata geological, even cosmological.

Till a' the seas gang dry, my dear,
And the rocks melt wi' the sun!
And I will luve thee still, my dear,
While the sands o' life shall run.

It's an amazing verse. Here's a humble poet, a ploughman by original occupation. He appears to be describing how the Earth will eventually end after vast aeons of time. Today's scientists believe that this will happen in another several billion years as our dying Sun expands lazily into expiration. In these lines Burns has the essence of it right, though he's more than a century ahead of the science. What's probably happening is less a case of scientific precognition than one of the bard melding ideas about the formation of rocks and the structure of time that were emerging during his own life and from the intellectual circles in which he moved. The story is a fascinating insight into the early development of Earth science and the kind of insights that have today evolved into Gaia theory. A key figure moving in Burns' circle was James Hutton, often called the 'Father of Modern Geology'. During the winter of 1786–87 Burns and Hutton met each other, along with the young Walter Scott and other luminaries of the Scottish Enlightenment, at the Edinburgh home of Adam Ferguson.[23] It was an occasion of sufficient social gravity for the painter C.M. Hardie to have captured it on canvas.[24]

As well as being a farmer, philosopher and chemist, Hutton was a physician by original training. His medical thesis at the University of Leiden had been called *De Sanguine et Circulatione Microcosmi*. But his real interest lay less in the circulation of blood in the microcosm of the human body than in the macrocosmic circulation of molten matter in the Earth. In pursuit of his field work Hutton had devoted thirty years of his spare time to riding around the countryside examining geological formations. He must have had a sense of humour, for after one particularly arduous day in the saddle in North Wales, presumably with his pockets bulging with freshly chipped-off specimens, he wrote: 'Lord, pity the arse that's clagged to a head that will hunt stones'!

In 1785, a year before the meeting with Burns, Hutton had presented his revolutionary conclusions about the origin of rocks to the Royal Society of Edinburgh. He called the paper and its subsequent book, *Theory of the Earth*. Up until this point in time most people had thought that sedimentary rocks were the residue of Noah's flood. But Hutton reckoned otherwise. He considered that geological processes are driven by the Earth's inner heat, much of it originally from the Sun, which melts and gives rise to igneous and metamorphic rocks. He saw that these forces – a precursor of the modern theory of plate tectonics – could account for the opening up of oceans and the raising of mountains. As a farmer he could appreciate that this is the mechanism by which rocks are constantly renewed to erode again back into soil, and that this sustains the very leaves by which we live. 'The globe of this earth,' Hutton said in his original 1785 treatise, 'is evidently made for man': it is 'not just a machine but also an *organised body as it has a regenerative power*'.[25] It is on this basis that James Lovelock credits Hutton's 'geophysiology' as a forerunner to Gaia theory – the idea that life creates the planetary conditions for its own stability.

In the eighteenth century many people still believed that the Earth was only a few thousand years old. Meticulous scriptural analysis carried out by the seventeenth-century Irish Anglican bishop James Ussher, had pinpointed the date of the Creation to the night preceding 23 October 4004 BC. But Hutton believed that the 'six days' of God's work were longer than what most previous human minds had imagined. He undertook his theology in *Mythos* rather than confusing categories of thought, as Bishop Ussher had done, with a misplaced application of *Logos*. The evidence that Hutton had gleaned while out and about on horseback pointed him towards what later became known as 'deep time'. As he was famously to put it in considering the aeons of the Earth, 'We find no vestige of a beginning, no prospect of an end'.[26] According to his friend Professor John Playfair, Hutton could describe all this with such eloquence that the minds of listeners 'seemed to grow giddy looking so far into the abyss of time'.[27]

Such is the scientific background to Robert Burns' 'Red, Red Rose'. In a classic demonstration of the Caledonian antisyzygy, Burns brings together opposites and holds them in alchemical tension. His poetry weaves the warp of science to the weft of the soul in a sublime synthesis. The 'two moods', the practical and the fantastic, *Logos* and *Mythos*, are threaded here to a seamless fabric by *Eros*. Here we have none of the late-Enlightenment narrow rationalism that only compounds the dissociation of sensibility. Here is the Scottish Enlightenment in its rare potential greatness. With an acuity like T.S. Eliot's bygone English metaphysical poets, Burns could see that the same 'deep time' that sustains the cosmos is the very crucible of deepest love. Such poetry is no mere play of clever words. Such poetry reflects the inner eye of the heart's perception.

In another of his poems, 'The Vision', Burns' eye perceives the Celtic goddess who is his muse. She has come, she tells him:

> To give my counsels all in one,
> Thy tuneful flame still careful fan:
> Preserve the dignity of Man,
>   With soul erect;
> And trust the Universal Plan
>   Will all protect.

She says that she had been with him from the womb: 'I mark'd thy embryo-tuneful flame, | Thy natal hour.' She watched as he grew through childhood and came to know 'the deep green-mantled earth . . . with boundless love.' She adored him notwithstanding all the foibles of youth, 'Misled by Fancy's meteor-ray. | By passion driven: | But yet the light that led astray | Was light from Heaven.' And it was from this deep sense of acceptance, from this sense of redemption held in deep time that Burns was able to pen 'My Luve is Like a Red, Red Rose'.

True love knows its roots to be eternal. That is why, in common with its manifestation in deep beauty, love always carries an element of sadness. Something in us knows that only beyond the constraints of this life, only outside of time, can love be fulfilled. True love is so special that we need to know it can be requited not just for a day, but as the lyrics of the 1968 Mary Hopkins hit put it, 'forever and a day'. This is the journey into the further reaches of human nature that the liminoid truncates. But the liminal, like the proverbial bridge over troubled waters, carries us across. If planetary hospice work is a fitting metaphor, then this undying love is the depth and beauty of the deliverance to which we are called. It is what renders troubled waters a wake-up call to the spiritual life. Only when we hit the narcissistic shit can we see the results of hubris, and that's all part of the growing pains that beckon us to heal the rift of inner and outer life. But to get to that point we have to find the courage with which to face Truth, and with it to face death as the seeming antithesis of life and love. We need the humility to ask the deep roots of life to provide the grace to gaze into the abysmal and yet, to hold fast and see beyond the despair. As a Buddhist monk once put it, 'The best place for meditation is in the jaws of the tiger.' That's where we find the conviction that Burns' muse gave him that our lives are deeply 'held'. That is how, slowly and falteringly, we reach that cosmic and yet very flesh-and-blood love that never lets us go.

The greatest of all the Irish harpists was Turlough O'Carolan. It was he, as Yeats reminded us, who received his music straight from out the faerie hill.

In Brian Keenan's historical novel, Carolan is lying on his deathbed.[28] Seated beside him is his lifelong confidante and possible lover, Mrs McDermott-Roe. She makes the following record in her diary. It sums up everything that we have been talking about in this chapter. 'Since I have been nursing Turlough,' she wrote, 'I find that the true face of death is infinitely loving.'

Such is the nursing of our planet to which we are called.

* * *

There can be times in life when consciousness transiently shifts and we glimpse realms of reality that manifests this love and the profound interconnection that it implies. For love is more than just the pull of two magnets brought close together. Love is structural. It is the underpinning of reality; the grand unification theory of the universe waiting simply to be embraced in the eye of Truth to be known.

For many years in my youth I made a study of paranormal

and mystical experiences. The two overlap but are not the same. The paranormal, or psychic, is about effects, while the mystical concerns the very ground of our being. But both add evidence to the idea that there is more to being human than secular materialism alone accounts for. There is an extensive literature, both populist and scientific, on all of this. For those of the latter inclination, I recommend a hefty tome, *The Varieties of Anomalous Experience,* published by the American Psychological Association. It is meticulously researched – rightly so to serve demanding intellectual standards – but dry as bones!

For those who prefer their bones to dance a bit, let me offer something different. Let me describe a personal experience. My interest at this stage lies in fleshing out the hospice metaphor of how we might respond to planetary crisis. My interest lies in seeking that quality of tenderness that Mrs McDermott-Roe found. I can't speak for my reader, but for myself I feel that if she could transcend the abysmal like that, then it testifies to a very special kind of strength. I'm not even sure from Keenan's text if the diary entries that he attributes to her were real, or a literary device that he made up. His presentation of them is ambiguous. I could try to find out, but I don't feel the need to. The poetic truth of the statement attributed to her is sufficiently compelling in its own right. That can be sufficient to carry us in the direction in which I invite my reader to accompany me.

My story is this. I also have another family account of similar calibre, but one will suffice for our needs here. During a period when I was home on holiday from university in the mid seventies, my father suffered a short period of illness. He had to spend some time away on the mainland as a patient in Raigmore Hospital. My mother had made plans to go and visit him. I can remember studying at my desk downstairs in the doctor's official house in Leurbost as I watched her drive off to Stornoway for the flight to Inverness.

My father had a whippet that he loved very much. Sheba had been rescued from ill treatment and was always very much 'his' dog. Usually she went everywhere with him – even snuggled under his tweed jacket on planes in those days before security got uptight. When he'd do hospital rounds he'd have her trotting around behind him. The patients loved it. It was a rural community, after all, and a dog – even a weird-looking one like a whippet – made the clinical surroundings more homely! But now it was different. As a patient himself in Raigmore he couldn't call the shots quite the same and so couldn't take the dog with him.

As I gazed out my window that summer's morning, some ten minutes after my mother's car had driven off to Stornoway airport, I watched in slow motion as a terrible thing happened. Sheba leapt from out of a hollow on the other side of the road. She ran over and died instantly as she hit, smack, into an oncoming vehicle.

I ran out of the house and reassured the distressed driver that he couldn't have done anything. Straight away I set about doing what a rural-raised child does in such a situation. I lifted up Sheba's limp body and carried her over to the cow's byre to get the spade. I wrapped her up in a hessian sack and solemnly processed the sorry bundle out onto the moor behind the house. There I sliced down through the heather turf. With tearful eyes I gouged out the peat to make a grave in which to lay Dad's dog to rest.

It was like I was on autopilot. At one level I was being utterly practical. At another, my whole being was awash with emotion. Looking back, I think that something in me regressed to a childlike state. Even though I was in my late teens I desperately wanted my mother to be there. I thought that if only the accident had happened ten minutes earlier, she'd have been able to carry the news in person. As it was, I would have to phone through and tell it to strangers – for there were no mobile phones in those days. It would be the hospital staff that would be giving Dad the news. It didn't feel like the right ordering of things.

I had just dug the grave to a suitable depth when I heard the sound of a car. To my astonishment, there was my mother reappearing at high speed over the brow of the hill!

I threw the spade down and ran over. She jerked to a halt and leapt out, distraught, with a haunted look on her face.

'What's the matter?' she shouted at me, almost accusatively.

'Sheba's dead.'

'Thank goodness,' she said as she took it in. 'I thought it might have been a child.'

My mother explained that she had been nearly seven miles away, just on the outskirts of Stornoway, when she became overwhelmed by a compulsion to turn back. She thought that something had happened to me or my sister. She was actually relieved to find that it was 'only' the dog. In those days she drove in to Stornoway several times a week. Never before had she experienced a compulsion to turn around like this.

She was able to make it back to Stornoway again just in time to catch her flight.

That afternoon she got to my father's bedside.

'I've got some bad news for you, Ian,' she said.

'I know,' he said calmly. 'It's Sheba, isn't it?'

The previous night he'd been repeatedly woken up by a vivid recurring dream. Sheba – the name means 'promise' – kept jumping up on the bed and pawing him.

A couple of years later, when I went to work in Papua New Guinea, I found that many other cultures accept this kind of experience as normal. People told me that telepathic or pre-cognitive dreams or apparitions were often how the village would get to know of the passing of a loved one far away. Neither is it only Papua New Guinea. Now that people know of my interest, I am astonished how often, especially when home on Lewis, people will tell me personal stories of the same genre.

It is one thing to have studied countless third party reports of such experiences but quite another to have one happen to oneself. To me, it adds to the evidence that profound interconnection – spirituality – really is for real and not just wishful thinking. Some people aren't moved by that sort of argument. They're either totally closed to any idea of spirituality or, at the opposite extreme, maintain that only 'faith' based belief from the heart counts for anything. But for me when I was at the age I was then, having an experience that blasted off some of the inhibiting intellectual cobwebs certainly helped to open my mind to spiritual possibility and to the underlying structure of the psyche.

Taking our argument deeper still, such phenomena can be a powerful challenge to the tendency that we all have towards selfishness. I suspect this is one reason why our culture generally keeps the door closed on intimations of the metaphysical. If there is strong evidence that our deepest being is interconnected at a psychic level to the rest of reality, then to pass by the suffering of the world must violate our very nature. No wonder if sensibility would then dissociate! Such is the price of selfish individualism that the Enlightenment thought opened the door to as it progressively stripped away the subjective reference points.

A spirituality of interconnection starts to heal this damage. Such is the structure of reality from within which, I believe, we need to start looking at climate change. This would carry us deeper than mere ethics and any notion of moral law. It would take us to the very source of conviction and motivation in the core of our being. As Arne Naess, the Norwegian philosopher of 'deep ecology', describes it: the human self ultimately participates in the 'ecological self'. At profound psychic levels, humanity is embedded in nature wild and free. That is what makes nature's beauty so restorative. Whatsoever we do or don't do unto this world comes back ultimately on ourselves.

* * *

And this music teacher . . . the one that I mentioned in Leurbost School . . . Duncan Morison was his name. For some reason we all called him 'The Major'. Like I said, he too had a beloved dog. A little white Highland terrier. He'd sit it atop the piano as he played. We kids were enraptured. Today there'd be a Health and Safety regulation that would have put a stop to it. Some child would get bitten by a flea or develop an allergy. Major would be sued. But not in those days. And I have this clear memory in my mind of one day putting up my hand. I had a question to ask. I can't have been more than seven years old because it was in Miss Montgomery's classroom.

'Please, sir,' I said, with a wry touch of boyish impudence.

'Yes, McIntosh?'

'Sir . . . why do we always have to sing songs about love?'

The point was that we boys didn't much like songs about love. We preferred gung-ho jingles – 'There was a soldier, a Scottish Soldier' or 'Campbelton Loch, I wish you were whisky'. We didn't quite vibe with what seemed to us to be the soppy sentimentality of Robert Burns and his red, red rose.

Quick as a flash the Major with the little white dog responded. It was with a speed sufficient both for my rebuke and to imprint his words forever on my mind.

'Because, my boy . . . love is the most beautiful thing in the world.'

* * *

We thought of him as our love child, this little life that Vérène and I were carrying in our marriage. And we spoke a lot together about 'sacred marriage' – as if there are three persons in a marriage, and the third is the source of life as love made manifest.

> As fair art thou, my bonie lass,
> So deep in luve am I,
> And I will luve thee still, my dear,
> Till a' the seas gang dry.

Only such depth of relationship can fearlessly face the shadow sides of life. This is the grounding in the Godspace. This is what sets all of life in perspective as it renews the worn and heals the broken. And when Vérène and I awoke the morning after Hogmanay, on New Year's Day 2007, our child was not moving.

Not realising the seriousness, I joked that maybe he'd had a night out on the town – a riotous Hogmanay. Maybe he was just having a long lie. But still he didn't move.

When the doctor turned on the scan I could see his little backbone. It was like we'd seen it before, but fully formed now. There was no pulsing from his heart. And I just assumed the doctor's sorry job for her. I quietly said to Vérène, 'He has left us, my dear. The bird has flown the nest.'

* * *

The witches told Macbeth he could not be harmed by any man of woman born. He thought himself safe from the consequences of his tyranny. But in the final showdown his adversary, Macduff, disclosed that he had been from his mother's womb 'untimely ripped'. And this was the expression that ran through my mind as the hospital staff conducted the caesarean. It was a difficult and risky procedure. There was a complication with the placenta. It caused the surgeons at Glasgow's Southern General Hospital to have to engage in a rare procedure. A considerable blood supply was on standby and an array of specialist staff in reserve. They were all amazing, and we were blessed to receive such care.

As the team operated, the words 'holy, holy, holy . . . glory, glory, glory' rampaged through my mind. Vérène's heartbeat monitor was starting to race. The procedure was difficult and the senior surgeon had to take over from his understudy. I had to stare into death: not just of our child, but also the small possibility that Vérène, too, could go. They had warned me that I might be asked to leave the theatre. Thankfully that, at least, was not called for.

Our love child was being born. In a vision in my mind's deranged eye I could see what looked like a wedge-shaped column of descending figures. At their head was a huge, solid, compassionate being with wide open arms. 'Vérène,' I whispered, as if with inner knowing. 'The ancestors are here to receive him into their number.' And the only place to go was onwards and inwards – deeper and deeper into the Godspace. These were the words that came to me:

ashes to ashes
dust to dust
life to life
love to love

And the midwife said, 'You're never unaffected, no matter how often you see it.' And after what seemed like ages she came round from the other side of the screen. 'It's a baby boy,' she said, 'And he's a beautiful baby. He's *really* beautiful.'

A little being like a teddy bear was laid on Vérène's breast. We wept without inhibition. We cried and cried because we loved him so much. How strange it was because I wouldn't have expected such an instant sense of connection and knowing. I felt like I would have given my own life for him. He was our love child. Our firstborn. Our son.

The doctors could find no reason why he should have died. These things just happen. He looked so perfect. His hands were like my own – broad and with long fingers. Huge feet, too. His face made me think of Vérène's the first time I ever saw her. It made me think of the wonder of a line that we'd read in a book together: 'The bond between a man and a woman is God himself, as is seen in the face of their child.'[29]

And we knew we would only know him again in God. And we knew we would only ever have known him in the deep time of God.

* * *

Two days later we wrote to friends:

> Our little son was born absolutely beautiful. We have named him Ossian Nicolas McIntosh. Ossian was an ancient Irish/Scots Gaelic bard who spent most of his life in fulfilment of love and art in the Celtic otherworld. His name means 'little deer' because his mother was turned into a deer and he was found on top of a mountain. We are pronouncing the name in a way that sounds like 'ocean' – *o-shee-an.*
>
> We are both astonished and heartbroken at the love we feel for this child. We have always seen our work with human ecology as being profoundly spiritual work, as it concerns the foundation of the human condition, and somehow we feel that his short and unborn life will carry that work forward in a world where so many people experience suffering.

I am sharing the intimacy of this story because I have spoken a lot about death in this book. It is not something into which to lead one's readers lightly. In our case we had expected to go home with a cradle. Instead it was a little white coffin. We can only say that our experience was that inner doors opened, and we were able to walk through without bitterness, even with joy. It was heartbreaking, most certainly, and yet blessed. It was a love that looks straight through and transcends death. And it felt like Ossian would always be spiritually with us – flying around the world, holding hands with us, like the little boy in Raymond Briggs' *Snowman* – wisely and laughingly present as we go about our work upon the Earth, right here, now.

* * *

It may not be possible for humankind to head off the consequences of the hubris that afflicts our planet. But if the worst comes to the worst, and if increased suffering falls upon life on Earth – then let us never forget that our spiritual imperative is to hold fast to hope, even where most grounds for optimism are lost. The words 'hope' and 'optimism' are often used interchangeably, but theologians see them as being very different. Jeremiah had no optimism that his people could be saved from falling into Babylonian captivity during the sixth century before Christ. In one of his Jeremiads he asks, 'How long will the land mourn, and the grass of every field wither? For the wickedness of those who live in it the animals and the birds are swept away.' And yet, even as the army of Nebuchadnezzar was laying siege to Jerusalem he put hope into action by going out and buying a field. He placed the title deeds 'in an earthenware jar, in order that they may last for a long time,' for he trusted that, one day, after the Babylonian exile was over, the land would return to the full glory of its potential richness. Ecological restoration would come hand in hand with social restitution. We could therefore probably get away with describing Jeremiah's field as having been the world's first nature reserve.[30]

The problem with mere optimism is that it's another example of the liminoid. It tries to alleviate suffering by denying reality. Hope, on the other hand, draws on inner resources that can co-exist even with outer pessimism or catastrophe. One can therefore be pessimistic about climate change but still retain hope. Indeed, without that hope we can forget about strategies for mitigation and adaptation. Only those resourced by hope – re-sourced by it – can circumvent the nihilism to which misplaced optimism otherwise turns. Alongside faith (which is to say, trusting perseverance), and charity (which is to say, love made manifest), hope is one of the three cardinal virtues of the spiritual life.[31] That is what makes it so important that public discourse and action on climate change is harnessed to a rekindling of the inner life. Only then can we face death's dark vale and get bearings with which to navigate Hell and High Water.

In the book *A Very Civil People*, John Lorne Campbell tells a wonderful story about three clergymen who came in the nineteenth century to try and proselytise the island of Mingulay. The youngest chanced to notice Ruairi MacNeil working barefoot in his field, digging with a spade. Anxious to impress his colleagues, this upstart evangelical from the mainland strutted over and demanded of the crofter:

'Do you know, my man, what Hell is?'

'Hell, my man,' said Ruairi: 'Hell is deep and difficult to measure, but if you keep on going, you'll find the bottom.'[32]

And that's the challenge that we face today with climate change. We have to peer into the unthinkable abyss. We have to be ready to have our prejudices confirmed, refuted and, more often than not, left churning uncertainly in the clammy sweat of something 'deep and difficult to measure'. The challenge of climate change takes us out of our normal constructs of reality. It pushes us to consider the options for creating a very different kind of world. In so doing, hope that 'if you keep on going, you'll find the bottom' becomes a veritable survival skill. I do believe that, come hell or high water, love deepens forever, and that is the meaning of this journey.

As such, modern climate change will be marked as a phase in geological evolution, but also, as a turning point in human consciousness. As global warming very likely confronts many with a metaphorical 'Babylonian exile', we too must work the Jeremiah's field imperative of placing trust in the possibility of reconstituting the world.

The artists always get it long before the intellectuals do. As Deep Purple sang in 'Child in Time', an epic heavy rock Jeremiad from 1970, we've crossed the line that distinguishes good from bad. In our blindness, we've fired the bullet of destiny. Now is the time to 'close your eyes and bow your head | and – wait – for – the ricochet'.

But for all that, for all the finger pointing and the justified guilt, there remains a paradoxical innocence. Maybe there was an inevitability that these things would happen. Maybe they are all part of an evolutionary journey of life on Earth that falls beyond the comprehension of conventional intellect and morality. For as the Deep Purple lyric suggests with tenderness that echoes Robert Burns, we remain, every single one of us, that 'sweet child in time'.[33]

And fare thee weel, my only luve!
And fare thee weel, a while!
And I will come again, my luve,
Tho it were ten thousand mile!

Chapter Nine

TOWARDS CULTURAL PSYCHOTHERAPY

In the 'Allegory of the Cave' that Plato tells in *The Republic*, Socrates invites us to 'liken our nature in its education and want of education' to a row of prisoners. He asks us to imagine that these have been chained up since childhood, their heads immobile and facing a blank wall deep inside a cave. Behind them is a walkway, and behind that, a blazing fire. People continuously traverse the walkway carrying statues of creatures and various objects of everyday life. The fire projects their shadows onto the wall and, for the prisoners, such a cinematic shadow play is their only known reality. But just imagine, says Socrates, that one of the prisoners is released. 'Then if he were forced to look at the light itself, would not his eyes ache, and would he not try to escape and turn back to things which he could look at, and think that they were really more distinct than the things shown him?'[1]

To cast light into dark corners is a work that is always fraught with danger. There's a Buddhist maxim that says, 'Do not disturb the sleeping'. It is much quoted by workshop junkies who seek a personal spirituality devoid of any political critique, especially if it touches wealth. I suspect that socially 'engaged Buddhists' would add to the maxim, '. . . until you're ready to hold the consequences!' Plato describes the consequences impeccably as his cave metaphor develops into a pre-computer critique of virtual reality. He has Socrates invite his listeners to imagine a further development. The freed prisoner is dragged up the passage, away from the fire and out into the world of full sunlight. Such is its dazzling brilliance that, at first, the erstwhile captive can see nothing at all – not even the shadows. Only gradually do his eyes adjust to the truth of reality. But then, adds Socrates in a final twist . . . imagine that he is taken back inside and chained down again onto his former seat where his colleagues pass their time placing bets on the antics of the phantasms before them: 'Would not his eyes be full of darkness because he had just come out of the sunlight? . . . Would not men laugh at him, and say that having gone up above he had come back with his sight ruined, so that it was not worth while even to try to go up? And do you not think that they would kill him who tried to release them and bear them up, if they could lay hands on him, and slay him?'[2]

This is why individual witness on its own, though important, is not sufficient. The changes needed to tackle the root causes of climate change must come about both individually and collectively. The one can only be sustained in step with the other. The starting point is to confess complicity in the problems and get beyond stage one – denial – in the planetary version of the Kübler–Ross grief cycle. Only then can our eyes collectively adjust to the sunlight and new options be appraised to address the problem of imprisonment. Only then will the current light-green shade of politics amongst political parties and environmental organisations alike ripen into a deeper fruition of meaning. How might that shift come about?

Aubrey Meyer of the Global Commons Institute sees this movement from tokenistic change to systemic change as requiring a programme of 'contraction and convergence'. The nations of the world need to contract their CO_2 emissions to a 'safe' level. But that can happen with justice for the poor only if nations progressively converge the differences that exist between high and low polluters. Ecological justice, therefore, cannot be separated from social justice. No nation must be able to score a competitive advantage over others by not embracing costly action. The general aim must be to work towards a world where the carbon budget of each country relates to its population.

To achieve this, nations would have to put aside naked competition and start moving in step with one another for the common good. It is why studying peace and learning how to cultivate trust is so important if there is to be any hope of tackling climate change at the roots. The process is a case of playing 'I will if you will'.

Attached to a wall in our house I have one of those little wooden toys – a 'climber' or *grimpeur* as it is called in the French alps, where Vérène bought it for me. You pull one string, and the first leg climbs a step. Pull the other, and the second leg goes up, this time higher. And so it goes on, one string after another, leg by leg, until the incremental climber gets to the ceiling. We pretend it's in the house for visiting children to play with, but really, it's for the adults – thus a little notice on the wall that says: 'Change comes each step in turn.'

Each of us, whether as nations or as individuals, can change a little in our lifestyle, but unless we imagine that we can stop the world and get off, there's only so far that any of us can go before having to wait for the surrounding society to catch up. We have to value self-restraint but it needs to be collectively applied. Slow, frustrating and inadequate though the pace is, lasting social change is an incremental process. The Buddhists advise that in such work we should 'seek the Middle Way'. It applies to social change as much as to personal ethics. They say: imagine you're on one side of a road in a moral issue. You want to cross to the other side but it's just too far to go. Be gentle on yourself. Confess your shortcomings but keep holding a mindful eye on the goal. Rather than try to go the whole way, aim only for the middle of the road. Once you get there and start adapting, that position may get more comfortable. You can then review where the Middle Way now lies. Its position will have redefined itself. Maybe you can take another few steps to the new middle of the road. Like the *grimpeur*, you can iteratively move closer and closer to the goal.

I like that. It works within inevitable human limits. It helps us to keep in check the doomed-to-failure egotism of getting ahead of ourselves.

Vérène and I find ourselves constantly challenged at a personal level by climate change. If we didn't confess our shortcomings, our many hypocrisies, it would make the whole issue impossible to talk about . . . which is probably why so many people don't much talk about it apart from projecting the need for action onto others. Like the prisoner from Socrates' cave, our eyes ache. We see, and yet don't like what we see, including our own complicity in the whole damned system! In some ways we live a relatively low-impact life yet we are hedged round with compromise, limited by daily life practicalities and, it has to be admitted, by personal weaknesses.

As a couple we have reduced our mileage by a third in the past four years, but still have a car. The food that we buy is mostly organic or Fair Trade, but I, especially, eat more meat and fish than my ecological footprint justifies, and cutting back so as to eat from lower down the food chain is one of the main ways in which most of us could reduce our environmental impact. Our household energy bill is below average, but that's mainly because we installed an ultra-modern virtually smokeless stove that allows us to turn off the central heating most evenings. We feed it partly with offcuts of untreated wood salvaged by one of my favourite pastimes – diving into builders' skips. But I can't be too virtuous about this. It only stacks up because hardly anybody else can be bothered with such seeming eccentricity. If everybody else joined the same ruse, we'd be having skip wars! I can therefore claim a niche solution using waste biomass, but one with only limited replicability.

Mindful of my carbon quota, I frequently turn down distant speaking invitations for one-off events, but some of my work involves long distances and I do still fly. I always have my justifications and sense of what is 'necessary', but the truth is that this is shaped by limitations on time, energy and money. Sometimes it has paradoxical effects. In the summer of 2007 I was speaking on the One Planet Leaders programme of WWF International in Geneva, and one of the people present was the

Chief Economist from IATA, the organisation that represents the world's airlines. It was just after a big climate change protest at Heathrow airport and the economist wanted to know what I thought of it because it rather concerned his line of business. 'I don't know,' I had to tell him. 'I wasn't there. I was flying out to France that day!'

What was so interesting is that the unexpected candour of the answer led to us having an inspiring discussion about what the airline industry could potentially do to reduce its environmental impact. This man clearly cared about the issue, though as he saw it, that care had to be held within the constraints of what his colleagues could accept. I suggested that a responsible approach would be to welcome proposals that aviation fuel should be taxed rather than to lobby against it. At present the industry enjoys exemption from duty as a throwback from the days when flight was in its infancy and governments wanted to get it airborne. That is what now has to change, and I suggested that IATA's role could be to endorse a transition that ensures an internationally level competitive playing field. The economist came over to me at breakfast the next morning, shook his head with one of those dismal looks that only members of his profession know how to pull, and said, 'I don't think that our Board will be ready to hear from you yet!' We both laughed, but at least the thought of pulling a wildcard from the pack seemed to have crossed his mind.[3]

Living as we all do in a fast-paced and interconnected world, the environmental cost of high-speed travel is particularly challenging to any of us whose lives do not involve staying at home. I have a friend who has made a pledge to try and never take the plane again. Her work for a government agency required attendance at an environmental conference in Italy. It took the best part of two days each way by rail from Scotland and cost three times the airfare. She feels she did the right thing, but will not be seeking to repeat the experience very often. Just how much her choice to reduce international involvement matters in the wider scheme of things depends on how vital her contribution is in that arena. That is not something that I can judge here. But as a general principle, if we want to have European neighbours that talk to one another and, most importantly, remain at peace, then we do need to have high levels of social connectivity. We also need more than video conferencing and e-mail, valuable though they are. We need flesh-and-blood interaction, too – the sharing of food, the subtleties of body language, the making of love.

That need for real presence is what I find creates the dilemmas. To travel by rail from London to, say, Geneva is a comfortable eight-hour journey on French trains and has everything to commend it. But add on the extra travel time from, say, Scotland or Ireland, or even a remote corner of rural England or Wales, and the edges of what can be personally sustained, if it has to be on a regular basis, become frayed. The truth is that green living requires that we slow down and that we do so in many different ways, but unless we try to drop out of our society and world issues – and that may or may not be a good way to promote change – there is often little option but to bite the bullet and run along with the herd. Sometimes it's the only hope of heading off the stampeding herd from the ravine.

We all fear being seen as hypocrites, but unless we openly confess our hypocrisy, then that fear – which is really just another ego thing – will hold us coddled in denial and unable to open our eyes or loosen our tongues. Of course, the churches are meant to specialise in confession and the framework of support that eases its way. In 1985 the Episcopal Church in America ran a wonderful proselytising ad that made the point. Printed over a huge picture of King Henry VIII was the caption: 'In the church started by the man who had six wives, forgiveness goes without saying.' That's probably the spirit that's needed for eco-confession, too. My own favourite example comes from the inspirational James Jones, Anglican Bishop of Liverpool. Here's what he said in an address that was delivered in St George's Chapel, Windsor, during 2007.

> I think it's important to preface any talk about the environment by a simple confession that we're all hypocrites. Very little human activity has no impact on the environment. I confess that I too have flown by budget airlines! Not so long ago I was

> clutching my easyJet ticket, dressed casually (not in a 'dog collar') and boarding a plane in Belfast to fly to the John Lennon International Airport in Liverpool. The storm clouds were gathering. We flew through the most almighty storm. Nobody spoke, nobody moved as we were thrown around the skies. On landing the man next to me began talking: 'I knew this would be a bad flight,' he said, 'but I didn't tell you as I didn't want to frighten you! Actually I'm an airline pilot. What do you do?' 'Actually,' I said, 'I'm a priest but I didn't tell you because I didn't want to frighten you!'[4]

We need the skills of pilot and priest. Both are equally scary in their own way. The 'pilot', as we saw in Part 1 of this book, is concerned with the technology, economics and politics of climate change. These are facets of outer life and between them they comprise one of the strings that cause the incremental *grimpeur* to start climbing. They frighten because the present-day technical possibilities don't add up to socially acceptable answers. The other string of the *grimpeur* is the realm of the 'priest' – the inner life and the insights it gives into the forces that drive consumerism. Only if the pilot and the priest both take steps in turn are we likely to be able to make progress towards the transformation of human society.

Critics can justifiably say that such incrementalism, as philosophers call it, 'will never get us there in time'. It's hard to argue with that. It's the justifiable criticism that's been made of the world's religions since kingdom come. And yet, what other models are there, short of revolutionary green totalitarianism? And I don't hear many people wanting to sign up for the RGT Party! It would entail violence that would be the very antithesis of a lasting solution. After all, Hitler was a vegetarian who had some wonderful green measures built into his fascist programme, but in the long term such association only damaged the environmental cause. The lesson is that if we want to achieve enduring change within democracy, we cannot set about it by trying to impose our will. That is why, in this book, I have emphasised the role of the metaphorical priest more than the pilot. Right now, that's the string that's most in need of the big pull. We need to reprogram, as it were, the software of the collective psyche, and not just tinker with hardware fixes. Yes, we should meet such-and-such a renewable energy target, but what is the point unless we also curb, and reverse, the inexorable rise in energy demand?

Global warming is but one of many presenting symptoms of the materialism that has placed us in this situation. Other symptoms include natural resource depletion, the degradation of the poor, the pollution and despoliation of beauty, discrimination against women and minorities, and always, underlying them all, the hydra's head of war. All can be derived from the formula that we explored earlier:

Hubris = pride   violence   ecocide

At the very root of hubris is the idea that 'man is the measure of all things'. In one way this is profoundly true. But it is only true when, as the ancients saw it and as religions like Hinduism and Christianity still teach, human nature is seen as being infused, at least potentially, with the divine. Without such a wider reference point, the 'Nimrod' within each one of us leads to psychic collapse into the vacuous shell of our own egocentricity. Addictions are nearly always attempts to mask that vacuity – alcohol, opiates, nicotine, and consumerism – pretty much in that order. But it's a counter-productive masking. As Kevin MacNeil describes things in his novel, *The Stornoway Way*: 'The bottle wasn't half full. It wasn't half empty. It was entirely empty. Drinking alcohol is like filling yourself with emptiness.'[5]

I believe that similar principles to those by which other addictions are tackled now need to be extended towards consumerism. For example, the experience of Alcoholics Anonymous is that addiction can only be conquered by confessing the problem and seeking help from a 'higher power'. It aims, as it were, at a reconnection of the human with the divine that involves the renowned '12-step' recovery programme. My question is therefore, 'What might a 12-step programme to counter the addiction of consumerism look like?' What steps might we and our societies consider to heal the dissociation of sensibility that has left so many struggling to be satiated?

A 12-Step Programme

The following preliminary steps are all elements of what I would call a 'cultural psychotherapy'. Just as psychotherapy with individuals usually tries to help people to understand their own history – what has made them how they are and how they are not – so cultural psychotherapy does the same at collective levels. It is what any psychologically aware teaching of social history ought to express, though it rarely happens because the implications can be explosive.

In *Soil and Soul* I attempted to illustrate cultural psychotherapy with land reform and community empowerment. It is not a case of having 'a therapist' set loose on an entire community or nation. Neither is it something easily written into the programmes of governments and institutions. It is something much more subtle and participative than that – more a process of osmosis that comes from many people and directions at once. It happens when any one of us contributes to a climate in which truth is spoken and a context for depth is held. I think there are three main elements. There is the *re-membering* of that which has been dismembered. A *re-visioning* of alternative ways in which things could be. And a *re-claiming* what is necessary to bring that vision to fruition.

Such an ethos can propagate in many different ways – through community meetings, in all manner of media, in art and music, in education, health and religion, and in grassroots organising. Rarely can it be led by the political mainstream. Most often it is a movement of the avant-garde – digging out the channels into which subsequent political process might flow. We saw this very clearly with Scottish land reform. What started off as a few visionary trickles from local communities in the early 1990s built into a river that eventually delivered the Land Reform (Scotland) Act 2003. Today, more than a third of a million acres representing 2% of the land area of Scotland are under the ownership of community land trusts, and a raft of revolutions in community regeneration has followed in the wake. People in government often say to me, 'What should we do about climate change? We can't disagree with the logic of the scenario you paint, and yet it makes our efforts seem futile.' I find that all I can answer is, 'Keep pushing the limits of what you can do within your electoral mandate, but above all, wake people up.' It is a tough answer but a true one.

Our difficulty in tackling global warming is that it is a symptom of malaise in the collective soul. Unless the psychospiritual roots of this are grasped, our best efforts will amount to no more than 'displacement activity' – like the wild animal that, powerless to defend itself from a predator, starts grooming itself. By collectively refusing to wake up and radically transform our ways of life, conflict and ecocide is likely to knock ever louder on the back door. Like King Duncan's horses breaking loose in *Macbeth* 'as they would make war on mankind', nature's horses of the apocalypse will not let us off the psychological hook. We are faced today with a collective neurosis. For as Jung wrote, 'People who know nothing about nature are of course neurotic, for they are not adapted to reality.'[6]

The steps that follow, then, are all about reconnecting with reality. They are an attempt to describe patterns and examples by which we might re-establish sensibility. I have written them up in 'we must' form. That way adds a bit of rhetorical oomph. But, of course, they are only suggestions, faltering and contestable at that.

1. *We must re-kindle the inner life*

This is where it all must start. The inner life is our most fundamental resource. It is the realm of thought, creativity, imagination, emotion, visions and dreams. It falls both within and beyond our conscious ken, starting in the individual mind but anchored to the eternal Spirit. We will not be able to live sustainably on Earth nor deepen human dignity unless we learn how to be resourced from such roots. I have suggested that violence historically destroyed much of our capacity for the inner life and has subsequently limited it from re-emerging. The antithesis of violence is empathy – felt connection in loving relationship. Rekindling the inner life is therefore about opening to empathy. That includes its expression through family, friends, community, work, the arts, nature and psychospiritual development.

Awakening others to the inner life is perhaps the most important contribution that the artist can make to present times. Rekindling it is a process that takes many years in most people. It happens in fits and starts, sometimes seeming to run more backwards than forwards. But gradually, steadfastness develops. A person with a well-developed inner life finds grounded strength within. This is what the violent can neither stand nor understand, but this is what sustains the world and living things.

Whether we are aware of it or not, spirituality is the powerhouse of the inner life. It is the inner reality or 'interiority' of all things, akin to the role of energy vis-à-vis matter in the outer world. Like any form of power, spirituality can be corrupted. Religious history is full of it – tormenting inquisitors, mad mullahs and paedophilic priests. But always the name of the game, as Walter Wink's writing so powerfully shows, is to call 'fallen' power back to its higher, God-given vocation.[7] The aim is not to destroy but to redeem. This is where spirituality is revealed as that which gives life and, specifically, life as love made manifest.

I believe that none of us can force love to happen. It doesn't come from an act of will mandated from the ego. Love is an opening, a gift of Grace. It comes from the Spirit that animates the soul, and is within conscious intent but beyond conscious control. We can ask to love and be loved, but usually we must wait. In the waiting we have to sit with our emptiness. That's where courage is called for. The courage to face the truth without resorting to the masks of lies and addiction. It is the deepest meaning of prayer.

God does not force love upon us. It's not a rape, in spite of the sorry impression that spiritual abuse by cults and some organised religion has all too often given. The reality is that God simply invites us to say 'yes'. All else follows on from that deep letting go into Being. It is how the inner life rekindles from its primal source, one that may often trickle, but which never runs dry.

2. *We must value children's primal integrity*

Our children are shaped partly by their intrinsic potential – both genetic and spiritual – and partly by the social and natural environments that surround them. When either of these are degraded or marred by violence, the child is at risk of becoming a product of a damaged world.

Each child is a seed waiting to flower into its own destiny. That seed is the child's primal integrity, its innermost soul. Caring for this means neither neglecting nor indulging the child. It means helping to birth its *areté* – its all round potential – across a range of competences that integrate head, heart and hand. It means, above all, communicating empathy by expressing respect and, equally, graciously accepting its reciprocation.

Psychologically speaking, the 'first half of life' is about developing an ego identity. Here we learn to wash our face, express what we're about, and make a living in the world. What distinguishes a child where the primal building blocks are well positioned is the ease with which transition can later be made into the 'second half of life'. This is the deepening into the soul – the realisation, as we saw earlier, that one is not just the cork but also integral to the river that carries it. Such inner-resourced adults make bad addictive consumers. Their sense of well-being comes mainly, though as corporeal beings never entirely, from things that cannot be bought or sold.

All children need safety and stability, social networks where they can make well-formed attachments in relationships, and contexts where they can express without fear what is happening in their lives.[8] The provision of these should be the cornerstone of public policy and family practice. Neither should attention to primary needs in one another cease as the first half of life matures. Children can remind us that such principles remain important even in the 'second childhood' of old age. That means honouring primal integrity all the way from the cradle to the terminal letting go that, one day, will signal the 'passing' of a life fulfilled.

3. *We must cultivate psychospiritual literacy*

Implicit to what has been said so far is a framework of understanding of what it means to be a human being. In the past, religion defined this with its creeds and dogmas. At their best, these express profound truths. But spiritual abuse within politicised religious structures has too often soured them. It has left many potential followers allergic, fuelling the rise of secular humanism since at least the Enlightenment.

The bridge between rationality and spirituality started to be rebuilt during the twentieth century by depth and, laterally, transpersonal psychology. Depth psychology was pioneered by Jung, and transpersonal psychology is its late twentieth-century flowering into a spiritual psychology that is built upon the psychic interconnection of all things.[9] Since the turn of the millennium, words like 'psyche' and 'spirituality' have become increasingly acceptable in mainstream public discourse. This has been helped along by many people now drawing a distinction between religion and spirituality.

At its best, religion is the socially organised structure of communally expressed spirituality. The religions of the world should be the culturally appropriate trellis up which the living vine of spirituality can grow. But where religion has become dysfunctional and the trellis no longer leads towards life, the vine is perfectly capable of growing wild. That is what we commonly see happening today. It is a healthy development – provided that sight is not lost of the fact that spirituality does concern the dynamics of interconnection. Community is therefore a key part of it, whether we name it 'Sangha' (Buddhist), 'Ummah' (Islamic) or 'Church' (Christian). Spirituality does require withdrawal and private retreat, but this must interplay with a social context. There can be no such thing as a wholly private or privatised spirituality.

In my own work speaking to many different types of group – environmental, church, corporate, military, governmental – I often find it useful to communicate a basic structure and terminology of the psyche. It is only a model and a simple one at that, but I find it invaluable for creating a shared starting-off point. This is what I mean by psychospiritual literacy. What I do is to hold up the back of my hand and say:

> The structure of a human being can be modelled like this. Here's a finger nail. That's my ego self – my small self which is the outer self that is Alastair McIntosh. It's centred in my field of consciousness. It's the me who's giving you this lecture, who has done this and that in life and, oh yes, hopes to flog you one of his books afterwards! That's ego for you! We've all got one and actually, we all need one. It's our face in the outer world and building it is the psychological task of the first half of life. It's like something that my friend Djinni of Scoraig did. Once she stood outside a potentially fraught community meeting holding a box marked *Ego*. As people arrived she enquired jauntily: 'Do you need some, or would you like to leave some of yours here?'
>
> Right at the base of my finger is the hand. Where the finger joins the hand is my deep Self, the great Self or the soul. The capitalisation there is deliberate to distinguish from small self. Here is the part of me that connects to the undercurrent of the Spirit, the animating fire that is 'God' within. The deep Self is the ultimate grounding of who I am; the deepest me and the crucible of inner life. It sits, at the boundary of time and the eternal, at the juncture of the personal and the collective unconscious. We are not normally aware of these realms, but notice that there's several spread fingers on this hand. They're one another. The deeper we go the closer they come. At the level of the collective unconscious – down at the bottom – they're all joined, like islands beneath the sea. That's the nature of mystical interconnection. It is the ultimate basis of community.
>
> But don't get your harps out yet! In the middle, right between the small and the great selves, is my finger's knuckle. That's my shadow self. It sits at the level of the personal unconscious – the realm that's specific to my life but of which I'm not very aware.

> The shadow is the flip side of the ego's light; it's the murderous Mr Hyde that gives the lie to the charming Dr Jekyll. The shadow complex is charged up with all the hurts going back to infancy, all the things I've done or have had done to me that I've repressed, but also, all that I could be but have never yet become. Really, I'd mostly rather pretend the shadow's not there. Unfortunately, if you ask my close colleagues or my wife, they'll tell you otherwise!
>
> The integration of these three layers of being is called 'self-realisation' or 'individuation'. The name of the game if we want to become not 'perfect', but whole, is for the ego self to settle down to being held in the deep Self. Psychologically this is the work of the second half of life. Some people start working on it as early as their teens and others might reach their seventies but still be no more than uncentred teenagers. What makes it tough is that this journey requires coming to terms with the shadow self. Psychospiritual development always requires facing the darkness. Anybody telling you that spiritual development is all positive vibes and sweetness and light hasn't yet faced their own shadow, and a shadow unacknowledged is a shadow that gets projected out onto the world: a shadow that hits out in what I call 'shadowstrike'. That's why psychologically naïve groups are always infighting. A shadow that is faced, on the other hand, becomes the lode from which gold is gleaned. It becomes the coal face of both inner and outer growth. Humbled in our own humanity and tenderised towards others, the fullness of who we are can be gradually realised.

I am aware that some Buddhists would take issue with the schema presented here. They would say that there is no self or Self, therefore rather than creating psychospiritual literacy I am compounding the delusion. I suspect that there are depths of mystery here that surpass understanding, and that even their concept of no-self would collapse when pushed far enough – perhaps into 'Buddha nature'! Let me just emphasise that the simplified Jungian model I have presented is just that – a model. I find it useful, but that doesn't make it one size to fit all shapes.

4. *We must expand our concept of consciousness*

As we have seen in our exploration of advertising, the spirituality of consciousness matters because its hijacking is nothing less than a dangerous theft of life. Theologically speaking, to have consciousness captured is to fall into the hands of 'false gods' – those of money, power, fashion and insatiable want. When marketing substitutes real needs with artificial wants, it becomes idolatry – it requires the sacrifice of our lives' efforts for ends that can never fully please.

Western psychology and philosophy presumes that consciousness, as Professor Hans Eysenck personally put it to me in 1975, 'is just an epiphenomenon of brain activity'. So far neurological research has failed to establish how such an 'epiphenomenon' comes about. Eastern philosophy would argue that it never will be established. That is because the East considers the brain to be an epiphenomenon of consciousness, or 'mind', rather than the other way round! It is as if the brain is the radio receiver, but consciousness is everything that goes on in the recording studio that makes the programmes we listen to. The brain, of course, regulates consciousness. It has been described as a 'reducing valve of cosmic consciousness', just as a radio set can regulate the volume or tone of what programme it plays. But to look for the source of consciousness within the brain is like trying to find a studio full of performers running around the printed circuits of a microchip.

The idea that consciousness has no intrinsic existence is the root of nihilism – the idea that everything is meaningless. One of the problems with nihilism is that it removes all ethical constraints on behaviour. It permits open house in the manipulation of consciousness and thereby feeds both the degradation of human dignity and mindless consumerism. It is to counter such dehumanisation that re-humanisation, in the form of spiritual practice that acts upon consciousness, lies at the heart of all great religions.

In outward form spiritual practice involves such activities as prayer, meditation, yoga, dance, poetry, study, work, engagement with nature, singing and sacrament.[10] What all these share in common is a requirement for 'presence' – the process of becoming mindful to that which is real as distinct from that which is 'virtual' or a facade. This stimulates values that are more than mere ethical choices. It opens realms of motivation driven from inner essence. In some religious outlooks such as Quakerism or Ignatian spirituality this is thought of as being 'moved' or 'led' by the Spirit.

There is no rocket science in all this. Sages have taught it for millennia. Yet the faculty of consciousness is the first casualty of hubris. Conversely, hubris cannot bear to be exposed by mindfulness. It cannot prosper in awakened *Homo sapiens*, the 'knowing human'. That is why violence, the adjutant of hubris, is described by such adjectives as 'senseless' and 'mindless'. Violence can only arise in mindless ignorance of reality, thus Hinduism, for example, attributes evil to *maya*, which is 'ignorance'.

The development of consciousness is the antithesis of violence. It connects us with the fullness of reality, as we have seen, through empathy, which is love. Such is the shift from the liminoid to the liminal – crossing the threshold that distinguishes deathly nihilism from life-giving Being.

5. *We must shift from violent to nonviolent security*

Psychological advances since the end of the Second World War have opened new insights that offer hope for how violence can be reduced. But these insights are emotionally challenging to those who persist with a punitive approach, and so they have yet to be adequately integrated into public policy. 'Violence,' says James Gilligan, former director of psychiatric services in the Massachusetts prison system, 'is the ultimate means of communicating the absence of love by the person inflicting the violence . . . The self cannot survive without love. The self starved of love dies. That is how violence can cause the death of the self even when it does not kill the body.'[11]

Because those who have been desensitised by violence will be predisposed to its perpetration, Gilligan describes violence as a 'social epidemic'. The late Brazilian Roman Catholic bishop Dom Hélder Câmara first popularised this idea in his classic 1971 text of liberation theology, *Spiral of Violence.*[12] He said that social violence starts with the level 1 violence – the primary violence of social injustice. This leads to secondary violence – rebellion by the oppressed. That in turn invokes tertiary violence – repression by the powerful. And that further impoverishes the state and so completes the spiral by feeding back into more primary violence.

The only antidote to the spiral of violence is the spiral of love. This is the power of nonviolence, not as a passive 'pacifism' but as vibrant 'truth force' or *satyagraha* as Gandhi called it. Nonviolence has played a major part in bringing liberation to India, Portugal, the Philippines, South Africa, countries of the former Soviet Union, minority groups such as black Americans and dozens, if not hundreds, of other examples.[13] Is it not time to study peace and not just war? That, at least, is what I've said over the past decade in addressing some 4,000 senior officers from nearly 100 countries who have been through the Advanced Command and Staff Course at Britain's foremost school of war – the Joint Services Command and Staff College. A couple of hours each year are now devoted to exploring such a point of view in the curriculum. We live in strange times that can offer strange openings.

In my experience the military generally believe that while war may be inevitable, and that is what they train for, it is neither a good nor a lasting answer. Many today from the British armed forces have experienced active service in Northern Ireland, Iraq and Afghanistan. When I describe Camara's 'Spiral of Violence' or Walter Wink's naming, unmasking and engaging the Powers, heads begin to nod. These men and women – people for whom I have developed a paradoxical admiration because they understand the meanings of service and community far better than most of their political masters – also see peace as their business. We don't disagree over dying for one's beliefs. Our point of contention is whether also to kill for them. The bottom line question is: 'wherein lies true security?' I simply suggest that we need to shift along the spectrum from violent to nonviolent forms of security. Climate change demonstrates this imperative better than anything. We're not going to head off global warming by continuing to bomb our way into other people's oilfields. The only hope is moving towards social and environmental justice across the world. Such has to be the cornerstone of an enlightened defence policy. It includes learning to recognise and process conflict at all levels of society.

6. *We must serve fundamental human needs*

As we have seen, the cutting edge of consumerism is the insatiable generation of wants. 'To be' becomes equated with 'to have' – what J.K. Galbraith called the 'dependence effect' of a cancerously corpulent economy. In contrast, a sustainable society, a sane society, is one that seeks to meet fundamental human needs in life-enhancing ways.

Such needs are called 'fundamental' because happiness only requires a certain level of materiality before the balance of fulfilment shifts from outer acquisition to the inner capacity for appreciation. A sane society would be one that satisfies fundamental needs for shelter, food, water, education, health-care and so on, but which also stimulates people onwards into realms of life that money cannot buy.

The Chilean economist Manfred Max-Neef has undertaken simple but life-giving work on this. He suggests that fundamental needs are the same the world over and throughout history.[14] They include the needs for subsistence, protection, affection, understanding, participation, creation, identity, recreation and freedom. In a matrix he analyses each of these for what they entail in terms of being, having, doing and 'interacting' (which is to say, the nature of the relationships incurred).

For example, the need for identity means being in a sense of belonging, esteem and confidence. It might mean having a place of home, language, religion, sexuality, occupation, values, a set of memories and social reference groups. In terms of doing, it entails growing up, learning, working, worshipping, playing and, one day, dying. And all this requires interaction in the private and social settings of everyday life, with family, friends, colleagues, any chosen gods and goddesses and, who knows, perhaps a few enemies!

Joining this to a similar concept that was pioneered in the South Pacific during the 1980s by my former colleague, Fr John Roughan, now Secretary to the Prime Minister in the Solomon Islands, I sometimes carry out an exercise with groups where I present Max-Neef's fundamental needs as an array of spokes on a wheel. Participants are invited to shade in, perhaps using a score out of ten, to show where they see themselves, their communities, or their nations on each parameter. In the South Pacific version this was called the 'Development Wheel'. It used indicators of village wellbeing like those of Max-Neef, but including practicalities such as water supply and food availability. It proved a crucial tool to rapidly appraise village relief priorities after the devastation of Cyclone Namu in 1986.

Whichever version of the method is used, once the segments of the wheel have been shaded in participants can be asked the pivotal question with which to reflect on the status of their needs: 'If this was a bicycle wheel, what sort of a ride would you have?' Some people have a frustratingly constrained ride because all the spokes or segments on their wheel scored low. Most have a bumpy one, indicating punctures in some areas or lack of growth. These then become the priority for attention. Rarely is anybody coasting full speed downhill!

Having evaluated their needs, participants can then be asked how they go about trying to satisfy each type of need. Max-Neef's view is that fundamental needs are shared by everybody. It is the way we try to satisfy them that differs culturally and from person to person.

He analyses 'satisfiers' of needs as follows. Some supposed satisfiers are 'violators' of others. For example, the arms race supposedly satisfies the need for protection, but it violates needs for subsistence, affection, participation and freedom. Others are 'pseudo-satisfiers'. For example, prostitution is only a surrogate for affection. Others are 'inhibiting satisfiers': television targets our need for recreation but can inhibit creativity. Some are 'singular satisfiers': they satisfy only one need rather than many, for example, food hand-outs that satisfy subsistence needs alone. And lastly, the ideal is to try and achieve 'synergic satisfiers'. These satisfy a wide range of needs in ways that stimulate the creation of whole networks of wellbeing. An example is how babies are fed. Bottle-feeding can be singular. It can satisfy little more than the need for subsistence. But breast feeding additionally usually satisfies needs for protection, affection and identity.

In such ways Max-Neef's model challenges the conventional idea that a society should maximise economic growth. Instead it shows how qualitative outer and inner aspects of consumption – being, having, doing and interacting – are interwoven. The sane society, he says, is one that would assess socio-economic policies to optimise the satisfaction of fundamental human needs synergistically.

7. *We must value mutuality over competition*

Without the competitive ethic, modern life would be a very sluggish affair – perhaps not unlike the former Soviet Union. Competition both motivates and challenges towards perfection. However, competition becomes a destructive force if not held within a wider framework that is cooperative. Today, obsessive competitiveness is pushed in government policy, industry and even at children in the classroom from the most tender age. In theory it does not have to be aggressive in this way. The original meaning of 'competition' derives from the Latin, *competere*, from *com* meaning 'together' and *petere* meaning 'to strive or seek'. *To compete* therefore originally meant, 'to strive in common'. But it is a sign of the times that this sense has largely been lost. Instead an expression of behaviour has evolved that has become injurious to the soul and destructive to the environment. People are encouraged to compete and consume not out of need, but to keep up – ever fearful that if they don't run faster and faster on the racetrack of success they'll be trampled by those coming up from behind.

The counterpoint to such competition is cooperation. At a commercial level this finds expression in cooperative and mutual business entities. But how can such cooperation be kept on its toes? Do we not need a bit of both qualities? Like Plato's image of the soul as a chariot drawn by two horses, one passionate and the other reasonable, could there be a higher synthesis by which such opposites can pull together?

I recall discussing this question with my friend Thierry Groussin, head of training in the French cooperative bank, Groupe Crédit Mutuel. We were driving around on the single-track roads of South Harris. Being the kind of place it is, as cars met in opposite directions, they'd typically pull in and flash one another to move ahead. Sometimes they'd cause mini traffic jams playing 'You go. No, you go!'

'There you are, Thierry,' I said. 'This is the community where people compete to cooperate!' And that's how the horses of competition and cooperation can be harnessed. Naked competition is based on naked individualism. We are, of course, individuals, but as we saw with the metaphor of the back of the hand, we are also interconnected with each other and therefore interdependent upon one another. As such, business structures based on mutuality reflect depth psychological reality better than do those based on aggressive competition. Perhaps there was method in the word's original meaning.

Fascinatingly and perhaps disturbingly, it is not just the green movement that is exploring this. Mainstream marketing is also flirting at the edges. Charles Saatchi is no longer involved with Saatchi and Saatchi. The company's new French owners have introduced an ethical policy that prevents the acceptance of accounts for products like tobacco and armaments. But as Europe's biggest ad agency, Saatchi and Saatchi remain at the cutting edge of marketing ideas. In a controversial book called *Lovemarks,* their worldwide CEO, Kevin Roberts, claims that 'love' is now the cornerstone of Saatchi's strategy! He says:

> Today the stakes have reached a new high. The social fabric is spread more thinly than ever. People are looking for new emotional connections. They are looking for what they can love . . . When I first suggested that Love was the way to transform business, grown CEOs blushed and slid down behind their annual accounts. But I kept at them. I knew it was Love that was missing. I knew that Love was the only way to ante up the emotional temperature and create the new kinds of relationships brands needed. I knew that Love was the only way business could respond to the rapid shift in control to consumers . . . The idealism of Love is the new realism of business. By building Respect and inspiring Love, business can move the world.[15]

We need be under no illusions that Kevin Roberts' primary loyalty is, as he implies, to his clients' brands. Neither need we imagine that upping the emotional ante in this way is going to cut consumption. But as his close colleagues have told me, 'It will change consumption.' And how very, very interesting it is to see that even the business world is starting to feel challenged by the need for a new relationality.

In early 2008, when I tested some of the ideas in this book with a futures think tank run by WWF-UK in London,[16] it was a senior executive from Saatchi's who seemed to be one of the most switched on to them. The greater part of me is, of course, suspicious. At the end of the day, it is hard to see how capitalism can survive without being cut-throat and without marketing forever inventing new tricks. But another part of me suspends judgement. There's a case for stopping and watching what happens at the passing place on that single-track road. Could it be that a 'new realism' based on love and respect is starting to become visible as a product of contradictions in the nature of advanced capitalism; an emergent property not previously imagined? Could that be what in future might shift our economic system unexpectedly towards mutuality? Could it thereby 'move the world' in the manner Kevin Roberts hints at? I will watch with interested scepticism and a potential helping hand, but we certainly live in peculiar times.

8. We must make more with less

One way that social and environmental justice movements have learned from the marketing world in recent years is in their understanding of product augmentation. This shows in the growing market for products with social or environmental 'kite marks' – Fair Trade for better prices to the poor, Soil Association for organic foods, RSPCA Freedom Foods for animal welfare, and so on. The added value accrues because values are built in. For example, as much as we can in our home, we buy certified organic meat from the local farmers' market. It costs double what we'd pay for the generic product in a supermarket so we eat smaller portions and a little less often than might otherwise have been the case. But it pleases us more, because we know it treats the soil, farm workers and the animals better. You're not left feeling tainted afterwards like you might with a leg of cheap imported battery chicken. As such, less becomes more. You don't just buy food. You buy something consistent with your understanding of right relationship.

One of the trade indicators that most gives me hope for the future is that, worldwide, the market for Fair Trade certified products grew 42% in 2006. It directly benefited more than 7 million producers.[17] To my mind, paying for things like that is better than giving to charity. It embodies justice and so upholds dignity. None of the genuine ethical products require coercion or manipulation to make them leap off the shelves. Demand simply comes from people's growth in consciousness; from a growing activation of the inner life.

To activate the inner life means to deepen the capacity for presence. Presence applied to what we consume means a savouring of things, a drawing out of the full satisfaction that something can give us because our attitude receives its totality. For example, when I hold in my hand a glass of good malt whisky I don't just gulp it down. At least, not on the first glass! I cradle it around and warm it with my hand. I'll spend several minutes before tasting a drop, just enjoying the aroma. It is the distillate of the land that I imbibe – the essence of barley, peat, the sea and our people who worked it. Those are my words, but the producers know it too. As one manufacturer says in the instructions that come with the bottle: 'Touch it. Feel it. Form a bond with this place and the people that live here. Become one of the many people bound to this place by their love of Laphroaig whisky and all that it embodies'!

My reader may think I've been taken in too much by the advertiser's motivational manipulation. I would dourly reply that such a reader cannot be a Scot! What we see here is the truth of the culture driving advertising rather than the other way round. But if we're going to have to argue about it, let's do so over a glass of . . . well, actually, Lagavulin if you could possibly stretch to it? I mean, we all have our foibles! But whatever the outcome of that debate (and I'm happy to stand corrected if we're still standing), my point remains that it is the combination of product and presence that builds augmented satisfaction. Whisky demonstrates the principle perfectly: the older the bottle, the smaller the dram. Applying the same principle to consumption as a whole, we are all called upon to become *connoisseurs*. That's how to make more out of less and be the richer for it. Far from being a recipe to kill joy, it's the only sustainable way to en-joy.

9. We must regenerate community of place

Ecology is the study of plant and animal communities in relation to their environments. Human ecology does the same with people. It studies human community in relation to its social and natural environments. Research in ecopsychology – ecological psychology – has repeatedly shown that we need to be able to attach to places as well as to other people.[18] We tend to be most at ease within ourselves when we have a sense of 'home' that is both emotional and geographical. As such, communities of place – our country, town, village or a bioregion such as an island or a watershed – tend to be very strong markers of identity. Bioregional identity is very often present in ways that we hardly notice. For example, when we speak of the 'Thames Valley Police' we are describing a bioregionally defined organisation. How strange that crime can follow ecology even if other walks of life can't!

Usually communities of place are stronger than socially constructed communities of interest. There is an asymmetry here. Communities of interest are nested within communities of place, not the other way round. That is because place is physical: it is our grounding in nature. Some postmodernists will challenge this. They say that nature, too, is a mere social construction. Well, at the cost of deconstructing any hope of a Lagavulin in the pub afterwards, I have an answer to that. This particular premodernist has a penchant for inviting extreme postmodernists to stop eating, hold their breath, and then we'll see how long their social construction of nature lasts!

'Place' is a very warm word. It is the product of both environment and culture; of nature and society. There is a sequence of reconnection with place that I have observed from my work with community regeneration in both Scotland and Papua New Guinea. I call it 'the Cycle of Belonging'. It functions like this:

1. A sense of place (grounding)
2. gives rise to a sense of identity (ego/head)
3. which carries with it a sense of values (soul/heart)
4. generating a sense of responsibility (action/hand)

That final sense of responsibility then feeds back into renewing sense of place. All this builds social and environmental cohesion. If the cycle is broken at any point, both human community and natural ecology are damaged – it becomes a vicious cycle. Conversely, if it is strengthened, people and place regenerate.

This has been the main dynamic by which community land ownership has achieved so much in Scotland. As Maggie Fyffe of the Isle of Eigg once told the BBC during a debate with the former landowner, 'In the past we never had the opportunity to prove that we could be responsible.' Some ten years after 'freedom' the island has a diversity of employment, a growing number of children in the school, and, since February 2008, its own community-run power grid – Eigg Electric – generating its very own 'Eiggtricity'! This is designed to provide over 95% of the island's requirements from renewables comprising three hydros, four wind turbines and an array of solar voltaic panels. Battery storage smoothes out supply and demand and diesel backup fills in when there's a shortfall. Previously most electricity needed by the island's forty-five homes, twenty businesses and six community buildings came from costly little diesel generators chugging away at the back of every home. Perhaps most interesting of all is that the new system works because load management is governed by a trip switch in every house. This ensures responsible awareness of what nature can provide: if anybody gets too greedy, they get cut off!

Strengthening people's connection to place can not only reduce greenhouse gas emissions. It also offers psychological benefits. I asked the husband of Eigg's doctor what changes he thought community land ownership had brought. He said his wife sees it the most. She prescribes far fewer antidepressants than was once the case! Another lesson from Eigg is that because the community buy-out was partly driven by conservation partnerships, and because tourism is so important, there is strong local support for the regeneration of woodlands and other special habitats. And lastly, when one goes into even some of the lowest income homes on Eigg, it is often Fair Trade tea or coffee that will be in the cup. Such can be the beneficial effects when community of place is re-membered, re-visioned and successfully re-claimed.

10. We must build strong but inclusive identities

The IPCC considers that by the middle of this century, 200 million people could be forced from their homes. These may lose all sense that they ever had of belonging to a place. In a cool, hilly country like Scotland where the impacts of climate change are likely to be less pronounced, pressure will grow to accept climate change refugees. What will happen? Will the privileged pull up the drawbridge to try and keep at bay the human consequences of their consumer profligacy? Or might we be able to think more humanely about who belongs to where? Specifically, can we bring to the cultural foreground constructs of identity that are inclusive rather than exclusive? Can we emphasise civic rather than ethnic identity: not 'blood and soil' but 'soil and soul'?

To answer this we might start by looking at what a nation is. I believe that a nation is more than just a state. A state is a mechanism of government, but a nation, over and above that, is a cultural entity. In a celebrated address at the Sorbonne in 1882, the great Breton theologian and Celtic scholar Ernest Renan pushed this even deeper. He said: 'A nation is a soul, a spiritual principle,' and he continued:

> A large aggregate of men, healthy in mind and warm of heart, creates the kind of moral conscience which we call a nation. So long as this moral consciousness gives proof of its strength by the sacrifices which demand the abdication of the individual to the advantage of the community, it is legitimate and has the right to exist.[19]

I find that very powerful. It affirms that identity, including national identity, is important. But it is only legitimate where it encodes a moral consciousness that can yield to admirable principles of community. Where people may lose their homes because of global warming, it begs the consideration that we who might be more fortunate have a special responsibility to take them in. Indeed, the legitimacy of our own claim to identity would depend upon it. It is for each nation to work out its own rationale, but in Scotland that might be built on existing cultural principles that define identity inclusively. For example, hospitality has traditionally been considered a 'sacred duty' for the short term and fostership, or adoption, for permanence. The fosterling must be protected, just as Joseph fostered and protected Jesus. As a Gaelic proverb says, 'The bonds of milk (i.e. nurture) are stronger than the bonds of blood (i.e. nature).' And another: 'Blood counts for twentyfold; fostership a hundredfold.' My own way of expressing this is that 'a person belongs inasmuch as they are willing to cherish, and be cherished, by this place and its peoples.'[20] The Scottish Government expresses it in the slogan: 'One Scotland; many cultures.'

None of these lofty principles mean that we are necessarily good at living them out. But it does help to have such cultural reference points in trying to do so. We, which is to say, a great many of us in modern Scotland during the national soul-searching that preceded Devolution, went out and looked for them.[21] They were waiting in our poets, our song writers, our customs and our history. Any nation could do the same if it so chooses. As Renan could see, it all depends what kind of a nation you want yours to be. What gives me joy in Scotland is that while there is still racism on the streets, there is also a passionate concern for the underdog, and this has very ancient anchor points. A practical example of it in action is the GalGael Trust in Govan where I live. The name dates back to the ninth century: the 'Gal' was the stranger, as expressed in place names like Galloway or Galway. The 'Gael' were the heartland peoples. Originally the Gall-Gael (*Gal Gaidheal* in Gaelic) referred to the 'strange' or 'foreign' Gaels – people who had interbred with migrants, mainly the Norse. Today's GalGael find this to be a powerful metaphor for present times. It helps to rekindle a strong sense of identity in people from hard-pressed communities even though many of us have very mixed backgrounds and fragmented identities.

Today there is a bit of the indigenous and a bit of the alien in most of us. A construct of identity that allows the honouring and melding of these can really work, and powerfully so, where community is at its heart.[22] The GalGael's workshops are constantly visited by politicians, clergy, academics and media interested in such an approach to community regeneration. It demonstrates that the Cycle of Belonging can apply to urban deprived areas as much as to rural ones.

A large part of GalGael's success is that participants are taught artisan skills using natural materials like wood, wool and stone. The boats that some of them help to build take them out on outings down the Clyde. The voyage doubles as a metaphor for the Hero's Journey of departure, initiation and return in life. People say things like, 'GalGael gave us back our river.' But more than that, the simple act of anchoring community in a context of mentoring and eldership gives people back themselves. It calls back the soul. Skills like these could be applied in many different contexts. They could, for example, help to ease the pain of cultural dislocation that people in the future are likely to experience from climate change.

As I struggle to put all this into words I am aware how difficult it is to communicate the sheer wonder of what is potentially open to us. There is something about the fullness of our potential humanity that can transfigure even situations that otherwise seem degraded, hopeless or pointless. In his essay, 'Real People in a Real Place' from a collection called *Towards the Human*, the late Isle of Lewis poet Iain Crichton Smith manages to touch on it. Here he reflects on a person who has moved to the city and been stripped of most of her anchor points. And yet, he manages to see beyond this into nothing less than the sacred:

> Sometimes when I walk the streets of Glasgow I see an old woman passing by, bowed down with shopping bags, and I ask myself: 'What force made this woman what she is? What is her history?' It is the holiness of the person we have lost, the holiness of life itself, the inexplicable mystery and wonder of it, its strangeness, its tenderness.[27]

Such is the depth from which we too must learn to understand identity if human potential is to be realised and a dwelling place furnished for all.

11. *We must educate for elementality*

In *A Sand County Almanac*, the classic work of ecology published in 1949, Aldo Leopold described conservation as 'a state of harmony' between people and the land. He proposed a 'land ethic' based on the principle that, 'A thing is right when it tends to preserve the integrity, stability, and beauty of the biotic community. It is wrong when it tends otherwise.'

The 'biotic community' is the whole web of life. People need to understand and be motivated to preserve it if they are to accept stringent action to combat climate change. In Britain since the 1990s considerable progress has been made in advancing environmental education. But education at the level of the 'head' alone is not enough. The head is very good at making decisions. But it is the passion of the heart that pumps blood both to the head and to the hand that puts action into effect. That is why Leopold's emphasis on beauty stands out.

At the end of the day, when the glitter of the shops has worn off, the packaging transmogrified to garbage, and the credit card bill slips through the door, the world of consumerism is sad and tawdry. Except where economic growth serves the fundamental needs of the poor, it measures little more than the rate at which natural beauty and human effort is trashed. Consumerism only goes skin deep, which is why it is always so obsessed with cosmetics, fashion and keeping up appearances. In contrast, true beauty is experienced when inner values harmonise with outer action. This is why right relationship with nature makes us whole. It salves our neurosis; it is a form of 'salvation'.

I am convinced, especially from my own experience growing up on the Isle of Lewis, that children both young and old need an 'elemental education' fully to be able to appreciate reality. They need contact with nature where they can learn about matter and energy, cosmology, the atmosphere and its weather, the soils and the rocks, and the rivers, lakes and seas and their flora and fauna. They need to experience nature's beauty and the sheer *fun* of it, for nature absorbs children in so many different ways.

We adults must be careful in our shouldering of the burden of awareness not to instil in our children the kind of eco-hypochondria that so often afflicts jaded greens – moaning about all that's wrong with the world, but forgetting to notice the magic of the crocuses pushing up and into blossom for yet another year. Children need to have a positive hands-on engagement with the ancient four 'roots' or 'elements' of reality – fire, air, earth and water. They need to know them in all their dangers – in their wild vicissitude that demands respect and courage on the Hero's Journey. And they also need to know them in all their sensitivity and vulnerability – in the filigree of frost on a winter's morning leaf – the hallowed loveliness that brings a tear to the eye.

In my view, none of this means treating the world in a hypersensitive 'precious' way. Even the most 'spiritual' of indigenous tribes kill animals, fell trees and hew stone. But it does mean doing these things with respect, with gratitude. We are bound up, all of us, with the strife of Heraclitus that constantly crucifies the elements of the world. And we, too, will feed the worms in our time. But these same elements are also bound into the love that, as Empedocles saw, unifies the world once more. We can see this cosmology symbolised in the Celtic cross. It is also the Medicine Wheel of Native American traditions from the other side of the great Atlantic. In these we have the four elements or the four directions. They're quartered by strife, but encircled in love. So there it all is – the elements of life incarnate dancing to the song celestial – life, death and resurrection.

12. *We must open to Grace and Truth*

At times I have been hard on organised religion in this book. And yet, whether because of or in spite of religion, I am conscious of having experienced a profound spirituality amongst the predominantly Presbyterian people with whom I grew up and move still to this day. I want to acknowledge my gratitude to them, for without their influence and especially their sense of community I could not have learned in the ways that I have. My father used to say that kindness is what matters above all else. Even when there is little else that can be done – when the actual or metaphorical floodwaters are rising all around and hope gets put to the test – even then, we can still try and be kind. Gratitude is what sustains and completes the cycle of grace. It is the essence of 'worship' – an Old English word meaning the celebration of 'worth'. If in our pride we neglect gratitude, or confuse it with sycophancy, there can be no hope of building true community because the doors of life's deepest gifts will stay closed.

The American environmental educator David Orr goes so far as to believe that gratitude is the single most important quality needed to address climate change. He says that only in such a spirit can we be freed from the loveless illusion of independence, and discover the sustaining truth of interdependence. This applies both for our relationships with one another and with the natural world. It is the flow of grace that opens the doors of 'providence', which is to say, *provide-ence,* in all walks of life. Such is what it means to find blessing. Orr quotes the great twentieth century rabbi Abraham Heschel, who said: 'As civilisation advances the sense of wonder almost necessarily declines . . . humankind will not perish for want of information; but only for want of *appreciation*.' And Orr concludes, 'In our universities we teach a thousand ways to criticise, analyse, dissect and deconstruct, but we offer very little guidance on the cultivation of gratitude – simply saying "Thank you."'[24]

Grace – both given to us and shared by us – walks hand in hand with another quality, Truth. The first verse of the Hindu gospel, the *Bhagavad Gita*, reads in Juan Mascaró's beautiful Penguin translation: 'On the field of Truth, on the battlefield of life, what came to pass, Sanjaya . . . ?' The battlefield of everyday life is here revealed playing out on a much greater stage – what is called Dharma in the Sanskrit or Truth in most English renditions, but meaning the unfolding through eternity of the divine cosmic way that fashions human affairs. Sanjaya was the eagle-eyed charioteer to the blind king, Dhritarashtra. The text's message is that power on its own is profoundly blind precisely because it finds Truth, and telling truths, so challenging.

Whether in the politics of a king or in the everyday lives of us all, power is always tempted to fabricate reality by putting a spin on things. As the great Russian philosopher Nikolai Berdyaev put it, 'The lie of the contemporary world' has come about because we have been subjected to or have permitted 'the disappearance of the very criterion of truth'. This distorts our perception of reality because it is 'the expression of a profound degeneration of the structure of consciousness.'[25] To restore consciousness to clarity and thereby, to redeem power in accordance with the Dharma, we need an external reference point that can lift us beyond our own restrictive narcissism. That is why we need the eagle eye of Sanjaya's spirituality to see where we're going. Church and state do not have to be allied, but if the actors of state lose touch with the truly spiritual they will slide inexorably into the lie.

We all get caught up so easily and deeply in webs of untruth and delusion. It makes truth and integrity one of the most challenging of spiritual practices. We perhaps fret, rightly, at our complicity in 'little white lies' but miss the structural whoppers. Unchecked, that makes it harder and harder to see Truth, or God for that matter. It is no coincidence that the secular age is the age of the lie, for consciousness itself has dimmed. Our sense of aliveness fades, and there's only the ache left behind – the lacuna in the soul – the promise of what could otherwise be that it's so tempting to try and attain through everyday addictions. As one of our community at the GalGael Trust told me, 'Heroin took away my pain, Alastair, but it also took away my soul.'

Soul retrieval is the ultimate ministry of the planetary hospice worker. That is the quickening – the polishing of the tarnished – that brings back life even in the midst of death. The more that I reflect on the culture of the lie in relation to what drives world problems like war and climate change, the more I'm convinced that the deep answer starts with trying to live truthfully. Along with gratitude, kindness, mindfulness and the love of beauty, Truth is the grace that kick-starts our lacklustre spirits back into touch. And so, we must re-set the little battlefield of our lives on the great field of Dharma. We can but consider simply saying 'yes'.

Towards the end of *Anna Karenina* there is a section where

Tolstoy's hero, Levin, is living 'in the very holy of holies of the people, the depths of the country'.

As he goes about the farm sowing, scything and threshing, only one other thing occupies his mind: the questions 'What am I? And where am I? And why am I here?'

He talks a lot with the muzhiks, the peasants, and Tolstoy tells us that 'all the good people close to him were believers.' During a conversation one of these contrasts two types of men.[26]

The first 'just stuffs his belly' and 'lives for his own needs'. The other 'lives for the soul' because, the muzhik says, he 'remembers God'.

'How's that?' asks Levin, who's desperately searching for the meaning of life.

'Everybody knows how,' replies the muzhik. 'By the truth, by God's way.'

And as Levin takes his leave, this man's words slowly come alive and set him ablaze. They have, Tolstoy narrates, 'the effect of an electric spark in his soul, suddenly transforming and uniting into one the whole swarm of disjointed, impotent, separate thoughts which had never ceased to occupy him.'

His hubris, his doubts and his existential angst are all dissipated. And so Levin lies down upon the good Earth. He gazes up at the cloudless sky and into the infinity beyond. He listens to mysterious, joyful voices from within and finds a depth of Truth, an unexpected gift of Grace, which transcends all the art and argument of the cynical world.

Here – beyond bounds of creed and dogma – is what the Church was always struggling to reach and teach.

Here, beyond mere belief, is faith – living Truth that quickens reality, renewing all that has been degraded by the familiarity that breeds contempt.

And Levin sees that 'reason could not discover love for the other, because it's unreasonable.'

He cannot believe his spreading sense of consummate happiness.

'My God,' he whispers, 'thank you!'

AFTERWORD

I have found this a terribly difficult book to write. *Soil and Soul* was such a good news story. Loads of people wrote in saying that it gave them hope. But with *Hell and High Water* it was difficult at first to see much hope. While undertaking the research I found myself spoiling several people's dinner parties and putting friendships on edge. It felt like I was poisoning debates with my pessimism. I was only speaking the truth as gleaned from combining scientific papers with political realism. But these left me doubting that humankind was going to be able to take the necessary steps 'in time' to head off climate change. It's said that a cynic is someone who has given up but not yet learned to shut up. Was that my problem? Should I really be writing this book?

It wasn't my idea in the first place to write about climate change. The idea had come from Hugh Andrew of Birlinn Ltd and he'd been encouraged to approach me by an old land reforming fellow soldier, Chris Ballance, who at that time was a Green Party member of the Scottish Parliament. Hugh wrote to me in the summer of 2006. He said he was after something on climate change that would ginger up debate for the Scottish parliamentary elections due in May 2007. He wanted something like a local text version of Al Gore's *An Inconvenient Truth* – short, snappy, and telling the politicians where we're at, where we need to get to, and how they can take us there. At that time I was struggling with the discomfort of seeing the glass half empty instead of half full. I hadn't yet dared to consider what might happen if we took it as completely empty!

I used George Monbiot's *Heat* as my starting point because I trust him. He'd danced at our wedding on Eigg, written the foreword for *Soil and Soul*, and we've sparked one another off since our South Pacific days. He'd written a book about Indonesian New Guinea while I was working in the independent eastern half of that island. If politicians want to examine detailed practical steps necessary to combat climate change, George's *Heat* is a good place to start. But he always knew that the remedy prescribed there would stick in the political gullet. 'Strangest of all,' he concluded, climate change 'is a campaign not just against other people, but also against ourselves.' And as he acknowledged in a subsequent newspaper article, people don't usually vote for austerity!

I spent the autumn of 2006 wondering what on earth I could say that could fit the policy development needs of a newly elected Scottish Parliament. Should I just summarise George and other writers like him but with a kilt around it? Should I reiterate policies already much mooted by the Greens which tend to be more optimistic than George is about the potential for solutions like micro-renewables? Somehow none of that quite fired me up. I was struggling with that emptying glass. Frankly, I was stuck for good enough answers. The carbon emissions implications of our energy demand and politically acceptable solutions just didn't square. And so I wrote back to Hugh Andrew and proposed a slightly different book. Could I do something that would touch on climate change, but would mainly emphasise community regeneration in the wake of land reform? That, at least, was 'good news' territory. It would make it easier to be optimistic.

Hugh was supportive, though we both knew that the aim was slightly off mark. Somehow, the whole paradigm of what needed to be said required shifting. The intended Scottish policy focus was important in terms of where I was coming from and the influences that have shaped my thought, but it was also a distraction. This needed to be a book that perhaps started in Scotland, but, like the wandering sailor, roamed the world before coming back home. Notwithstanding all these doubts, I was getting there. The research was on target, the first chapters had been drafted, and I hoped to have a manuscript sent to Hugh by his February deadline.

And then, on New Year's Day 2007, Ossian left. And what opened out in my demented mind was the timeless, mythological, bardic Ossian of our peoples. It was so strange choosing that name. It had come to me early on. Vérène liked it but said the translation would be problematic in French – it would sound too feminine for a boy. My son, Adam, lives in a tree house. Sitting with him one night round a stove high up in the branches, he said in a voice that seemed to come from somewhere else: 'That's a beautiful name. That's his name, but if you don't use it, I'll use it for my son.'

But the bird flew the nest. Madly I told Hugh that I would honour my obligation to deliver for February 28th. Wisely he excused me. But in the weeks that followed, the whole configuration of the book started to shift. It was death's empty glass that did it. Up until that point my subtitle had been 'Climate Change and the Human Condition'. Henceforth, and after the dust had literally been given time to settle back into the good Earth, it changed to: 'Climate Change, Hope and the Human Condition'. How very strange that it took a stillborn child to insert that word! How very special.

So, what has been achieved? I must concede that I have utterly failed to respond to George Monbiot's political challenge, at least, not in terms of any sort of conventional political argument. Let me try and explain why. Part 1 of this book, dealing with the science and politics of climate change, is merely a warm-up. It is a Trojan horse; for as the reader will by now have seen, the real writing begins in Part 2. There's a reason for this. Parts 1 and 2 represent the two worlds – the outer and the inner. They interpenetrate and yet there's a dividing veil. In the Celtic mythology of both Ireland and Scotland, Ossian went to live with the faerie woman, Niamh of the Golden Hair. After 300 years in Tir na nÓg, the land of eternal youth, he was pining for his old clansmen and wanted to pay them a visit. Niamh lent him her white horse, Embarr, but warned her lover never to dismount. If he did so his years would catch him up. Back in the world of men, Ossian was dismayed to find the great halls ruined, his friends all dead and gone. As he roamed the bleak transformed countryside he came across a group of men trying to lift a stone onto a wagon. He reached down to offer a helping hand, but as he did so, Embarr's girth snapped under the strain. Ossian tumbled to the ground. The years raced through him, his body took their terrible toll, and he slowly crumbled into dust.

That sense of working with two worlds is what it's felt like with this book. I have attempted to shift the paradigm of climate change over a threshold and into a realm that speaks in poetry, myth and vision. I have attempted a restoration to consciousness of the metaphysical that was crushed in the heartless vice of uni-dimensional positivism. Exactly how that translates back again into the world of politics, economics and technology I'm not quite sure. I think I can see the general principles. These are laid out in the 12-step programme in the last chapter. But to be any more specific than that would feel like setting hand to stone too firmly. Anything more than a light touch between the worlds risks snapping Embarr's girth and crumbling all to dust. It would be like pulling up the seedling by the roots every so often to check on how it was growing. I hope that this is not just evasion of the practical by transcendental displacement. Instead, I think that what has been laid out here – the imperative of rekindling the inner life alongside outer life – is an agenda that just cannot be pinned down in detail. It cannot be measured, controlled and managed with the boxes of all due performance indicators ticked. What is offered here is something complementary but different: a politics more of poetry than of prose.

But none of that is to discount the importance also of the prose! As I have been writing, the scientific indicators suggestive of climate change have relentlessly hurtled in. Some of what I've said in Part 1 will quickly be superseded by new research but the underlying principles will probably remain valid for some time. Most of Part 2, I suspect, will date no more quickly than the ancient sources on which I drew.

An example of how quickly things are changing in climate science is what happened in the Arctic during 2007. Early that year the scientific consensus seemed to be that the ever-mysterious Northwest Passage might be clear of summer ice within two or three decades. But by September, the European Space Agency had announced that the Passage had become completely clear for the first time since records began. The ESA described the rate of melting as 'extreme'. Satellite pictures revealed that a colossal 1 million square kilometres of sea ice had disappeared in comparison with 2006.[1]

I have been careful in this book not to be alarmist. Even my disturbing material about epidemics is carefully anchored to sources like the WHO, and many will say that I have not been alarmist enough. Yet at times, one cannot help but hear the thundering hooves and feel the hot breath of the apocalypse cantering by. There is a slow urgency about what humankind's levels of consumption are doing to the Earth. Slow, in that it is difficult to register it in ways that can trigger radical action on political timescales. But urgent, in that the ratchet is tightening especially if we have any care at all for future generations. The tipping points of no return show signs of slipping. That is why I have been forced to abandon optimism and seek recourse in hope.

My call to rekindle inner life is the next pull that's needed on the string of the *grimpeur* – the incremental climber. But that doesn't mean that we should wallow there! Too much inner life without the grounding nourishment of getting our hands dirty is just as toxic to the soul as the other way round. We need a dance between the fantastical and the practical; not apartheid between the two. Our drift must be towards becoming whole people in a whole world. We are talking here of a spirituality that is both transcendent and immanent. An incarnate spirituality that is not of the 'world' in all its wicked ways, yet neither abandons that world: 'For God so loved the world . . .' Even if we find ourselves forced to view the little steps we take today as patterns and examples for reconstruction after the grand melt-down in some post-apocalyptic scenario, what matters is that we never give up. Love does not succumb to compassion fatigue. Love cherishes the flesh-and-blood body of the world . . . infuses it . . . forgives it . . . constantly seeks to transfigure and to re-set the seeds of Eden as Heaven on Earth. That is our calling in these our troubled times. We are commissioned to draw out the flavour of what Providence provides; to be, in the words of the Master, 'the salt of the Earth'.[2]

And so, hope is not about sitting back on tenterhooks and waiting for a miracle to happen. Hope is being receptive to a new mind and a new heart. Hope is about setting in place the preconditions that might reconstitute life, and then getting on with it. All else is hubris on the bonfire of vanities. And I'm afraid that's as far as I can take this unfinished and unfinishable book. The rest is 'not perfected' because the rest is up to us all.

Such is the cry of the Earth to its own sweet child in time.

NOTES

A Note on Referencing

Much of the scientific material in this book came from up-to-date reports on the web. I have not printed the URLs here because they are long, unsightly, and highly prone to transcription errors, but live links to them all, valid as of early spring 2008, will be found on the *Hell and High Water* section of my website at www.AlastairMcIntosh.com/hellandhighwater.htm. I hope this will make things easy for my readers. Articles that I have not found on the web are referenced below according to standard conventions.

Please note that in a book of this range and complexity, some technical errors (as distinct from inevitable outdatedness) are bound to have crept in. I'm sorry if that's the case, but if readers would be kind enough to notify me of gremlins I will promptly post corrections to an erratum page that will be maintained on the said website.

Introduction

1. *Stornoway Gazette* (28 Dec 2006).
2. 'New Orleans recovery could take 25 years, Bush administration', *USA Today* (30 Mar 2006).
3. Joel K. Bourne, 'New Orleans' Rebuilt Levees "Riddled With Flaws"', *National Geographic* (6 May 2007).

4. Spencer Weart, 'The Discovery of Global Warming' (Jan 2008).
5. The National Center for Atmospheric Research, 'Frequency of Atlantic hurricanes doubled over last century, climate change suspected' (29 Jul 2007).
6. Scotland and Northern Ireland Forum for Environmental Research (SNIFFER), 'An online handbook of climate trends across Scotland – temperature related variables' (Met Office, 2006).
7. Intergovernmental Panel on Climate Change (IPCC), *Fourth Assessment Report, AR4 SYR Summary for Policy Makers*, 'Observed changes in climate and their effects' (Cambridge University Press, Cambridge, 2007). (NB: AR4 = *Assessment Report 4*; SYR = *Synthesis Report*).
8. George Monbiot, *Heat: How to Stop the Planet Burning* (Allen Lane, London, 2006), p. 16.
9. Manchester University Tyndall Centre for Climate Change Research, 'Living within a carbon budget' (July 2006).
10. IPCC, *AR4 SYR*, 'Observed changes in climate and their effects', Table SPM.6.
11. Monbiot, *Heat*, pp. xiv, 212, 215.
12. Leonard Cohen, 'The Future', *The Future*, Sony, 1992.

Chapter 1 – Nullius in Verba

1. This particular link is no longer available; however, the Royal Society website is a very useful starting point for the discussion of climate change. See the Climate Change page under their Science Issues link.
2. Alasdair MacIntyre, *Whose Justice? Which Rationality?* (Duckworth, London, 1988).
3. A.J. Ayer, *Language, Truth and Logic* (Penguin, Harmondsworth, 1961), p. 61.
4. See www.CharlesinSpace.com.
5. Richard Dawkins, *The God Delusion* (Black Swan, London, 2007), p. 57.
6. The Royal Society, 'A guide to facts and fictions about climate change' (March 2005). Also, 'Climate Change Controversies: a simple guide' (April 2007).

7. Channel 4 TV, 'The Great Global Warming Swindle', broadcast 8 March 2007, 9 pm. The quoted material was here on 30 April 2007 but has since been changed. See www.channel4.com/science/microsites/G/great_global_warming_swindle/index.html.
8. James Lovelock, *The Revenge of Gaia* (Allen Lane, London, 2006), pp. 5, 25, my emphasis.
9. J.R. Petit, J. Jouzel, D. Raynaud et al., 'Climate and atmospheric history of the past 420,000 years from the Vostok ice core, Antarctica', *Nature* 399, pp. 429–36. Press Release (3 June 1999): www.cnrs.fr/cw/en/pres/compress/mist030699.html.
10. NOAA, 'Trends in Atmospheric Carbon Dioxide – Mauna Loa' (2007).
11. IPCC, *AR4 SYR*.
12. Ibid.
13. 'Christopher Booker's Notebook', *The Sunday Telegraph* (11 Mar 2007).
14. George Monbiot, 'The Revolution has been Televised' *The Guardian* (18 Dec 1997).
15. Joanne Oatts, '"Global Warming Swindle" sparks debate', Digital Spy.co.uk (15 Mar 2007).
16. One of the first scientific critiques of the programme was to be found online, with commentary, at Real Climate: William Connolley and Gavin Schmidt, 'Swindled' (9 Mar 2007).
17. Steve Connor, 'The real global warming swindle', *The Independent* (14 Mar 2007).
18. Chris Merchant, 'Why the C4 documentary "The Great Global Warming Swindle" is wrong' (31 May 2007).
19. Richard Black, '"No sun link" to climate change', BBC online (10 Jul 2007).
20. *Crisis Forum* email list posting of 9 Mar 2007, also www.climatedenial.org on 3 May 2007.
21. Letter from Bob Ward, Senior Manager, Policy Communication for The Royal Society to Nick Thomas, Esso/ExxonMobil's UK Director of Corporate Affairs (4 Sep 2006) found online at *The Guardian* image file site.

22. Geoffrey Lean, 'Climate change: An inconvenient truth . . . for C4', *The Independent* (11 Mar 2007); Carl Wunsch, 'Swindled: Carl Wunsch responds' at realclimate.org (12 Mar 2007).
23. George Monbiot, 'There is climate change censorship – and it's the deniers who dish it out', *The Guardian* (10 Apr 2007).
24. Brendan O'Neill, 'Apocalypse my Arse', *Spiked* (9 Mar 2007).
25. Armand Leroi, 'Correspondence with Armand Leroi, Martin Durkin and others' (9 Mar 2007). NB: I emailed Martin Durkin on 30 April 2007.
26. Paul Driessen, *Eco-Imperialism: Green Power, Black Death* (Academic Foundation, 2006). Quote can be found at www.academicfoundation.com/n_detail/eco-imp.asp.
27. Paul Driessen, 'Eco-Imperialism chapter excerpts' (2006) found online at www.eco-imperialism.com/Corporate_Social_Responsibility_-_Chapter_Excerpts.pdf

Chapter 2 – Beyond Tipping Point

1. George Elder Davie, *The Democratic Intellect: Scotland and her Universities in the Nineteenth Century*, (Edinburgh University Press, Edinburgh, 1999).
2. Joseph Romm, *Hell and High Water: Global Warming – the Solution and the Politics and What We Should Do* (William Morrow, New York, 2007), pp. 154–55.
3. Shaoni Bhattacharya, 'European heatwave caused 35,000 deaths', *New Scientist* (10 Oct 2003).
4. David Adam, 'Does global warming kill 150,000 people a year?', *The Guardian* (19 May 2005).
5. John A. Church and Neil J. White, 'A 20th century acceleration in global sea-level rise', *Geophysical Research Letters*, 33 LO1602 (2006).
6. Catherine Brahic, 'Sea level rise outpacing key predictions', *New Scientist* (1 Feb 2007).
7. 'Survey: Glaciers in west China shrink 7 to 18% in five years', Xinhua (14 Dec 2007).

8. J. Hansen, Mki. Sato, P. Kharecha, G. Russell, D.W. Lea, M. Siddall, 'Climate change and trace gases', *Phil. Trans. Royal. Soc. A,* 365, pp. 1925–1954 (2007).
9. Romm, *Hell and High Water*, pp. 20, 79.
10. Alan Watson, 'Adapt and Survive?' *Ecos: a Review of Conservation*, Vol. 28:3/4, 2007, pp. 27–32.
11. 'Humans "affect global rainfall"', BBC online (23 Jul 2007).
12. J.R. Minkel, 'Darfur Dead Much Higher than Commonly Reported', *Scientific American* (15 Sep 2006).
13. United Nations Environment Programme (UNEP), *Sudan: Post Conflict Environmental Assessment* (UNEP, 2007).
14. Peter M. Vitousek, Paul R. Ehrlich, Anne H. Ehrlich, Pamela A. Matson, 'Human Appropriation of the Products of Photosynthesis', BioScience 36 (June 1986), pp. 368–373; Karlheinz Erb et al, 'Global Human Appropriation of the Products of Photosynthesis in 2000', University of Vienna (Oct 2006).
15. Dave Favis-Mortlock, www.soilerosion.net/, accessed May 2007.
16. Natural Environment Research Council (NERC), *The Oceans: scientific certainties and uncertainties* (Swindon, 2007).
17. Lovelock, *The Revenge of Gaia*, pp. 26–35.
18. Kenneth F. Drinkwater, 'The Response of Atlantic Cod (*Gadus morhua*) to Future Climate Change', *ICES Journal of Marine Science*, 62, pp. 1327–1337 (2005).
19. HM Treasury, 'Stern Review on the Economics of Climate Change' (30 Oct 2006).
20. Although an argument can be made that the Stern Review inadvertently does just that: George Monbiot, 'Juggle a few of these numbers, and it makes economic sense to kill people', *The Guardian* (19 Feb 2008).
21. See further discussion and references in my *Soil and Soul* (Aurum Press, London, 2001), pp. 37–9.
22. Camille Parmesan and Gary Yohe, 'A globally coherent fingerprint of climate change impacts across natural systems', *Nature* 421, pp. 37–42 (2003).
23. NERC, *The Oceans*.

24. Gabriel J. Bowen et al., 'A humid climate state during the Palaeocene/Eocene thermal maximum', *Nature* 432, pp. 495–99 (Nov 2004).
25. W.B. Yeats, 'The Second Coming', *Michael Robartes and the Dancer* (Kessinger Publishing Co, Whitefish, MT, 2003; facsimile of 1920 original), pp. 19–20.
26. NERC, *The Oceans*.
27. Walter C. Oechel et al., 'Recent change of Arctic tundra ecosystems from a net carbon dioxide sink to a source', *Nature* 361, pp. 520–523 (1993).
28. William Dillon, 'Gas (Methane) Hydrates – A New Frontier', US Geological Survey (Sep 1992).
29. 'Russia plants flag under N[orth] Pole', BBC online (2 Aug 2007).
30. 'Russia schemes to claim North Pole oil, gas, gold', Xinhua News Agency (2 Aug 2007).
31. 'Canada to strengthen Arctic claim', BBC online (10 Aug 2007).
32. Scottish Government, 'Scotland's Climate Change Programme: Annual Report 2007' (Mar 2007).
33. Andrew Kerr, Simon Shackley, Ronnie Milne and Simon Allen, *Climate Change: Scottish Implications Scoping Study*, Scottish Executive Central Research Unit (HMSO, Edinburgh, 1999).
34. Scottish Environment Protection Agency (SEPA), *State of Scotland's Environment 2006* (2006).
35. Ibid.
36. 'Extinction threat to Scots bird', BBC online (15 Jan 2008).
37. NERC, *The Oceans*.
38. Lovelock, *The Revenge of Gaia*, p. 55.
39. Dave Morris of The Ramblers Association Scotland, letter in *The Herald*, 29 January 2008.
40. Scottish Government, 'The Dancing Ladies of Gigha' (18 Sep 2007).
41. Peter Taylor, 'Energy Watch', *Ecos*, 28:2, pp. 82–87.
42. '"Zero carbon" homes plan unveiled', BBC online (13 Dec 2006).
43. Ibid.

Chapter 3 – Devil's Dilemmas

1. Proponents include Jeremy Leggett, C.J. Campbell, Paul Mobbs and Richard Heinberg. See, for example, Richard Heinberg, 'The View from Oil's Peak' (Aug 2007).
2. 'UK Fuel Tax: the Facts', BBC online (21 Sep 2000).
3. Jon Hughes and Mark Anslow, 'Power On,' *The Ecologist* (November 2007), pp. 35–47.
4. Monica De Wit, 'Hardwiring and softwiring corporate responsibility', *It's All Our Business: Corporate Responsibility in a Global World*, INSEAD Alumni Sustainability Roundtable (2008), 1:3.
5. National Petroleum Council, *Facing the Hard Truths About Energy: A comprehensive view to 2030 of global oil and natural gas* (NPC, Washington, DC, 2007).
6. Stephan Harding pers. com. from discussion with Richard Heinberg.
7. 'easyJet supports green air taxes', BBC online (18 Sep 2007).
8. Danish Energy Authority, *Energy Statistics 2006* (DEA, Copenhagen, 2007).
9. 'Natural Gas and the Environment', NaturalGas.org (2004). This data is taken from *EIA – Natural Gas Issues and Trends* (1998).
10. Alice Bows et al, *Living within a carbon budget* (Manchester University Tyndall Centre for Climate Change Research, 2006), p. 162.
11. See the community's website: www.scoraig.com and http://en.wikipedia.org/wiki/Scoraig.
12. Wikipedia, 'Freetown Christiania' (18 Sep 2007).
13. Alastair McIntosh and Michel Picard, 'Who is Your Enemy? Lafarge, NGOs and the Harris Superquarry Campaign' in *It's All Our Business: Corporate Responsibility in a Global World*, INSEAD Alumni Roundtable (2008).
14. Sustainable Development Commission, 'Tidal Power' (2007).
15. 'Climate Alarmists Consider "The Geritol Solution"', NewsMax.com (16 Mar 2007).

16. Donald Macleod, 'Footnotes', *West Highland Free Press* (1 Feb 2008), p. 10.
17. Energy Strategy and International Unit, Department of Business, Enterprise and Regulatory Reform (BERR), *Energy Trends*, (National Statistics publication, Jan 2008).
18. Posiva, 'Nuclear Waste Management' (4 Feb 2008).
19. Michael Settle, 'Total cost of closing down nuclear sites rises to £73bn', *The Herald* (30 Jan 2008); DTI and BERR, *Energy Consumption in the UK* (National Statistics, 2002).
20. Discussion in Monbiot, *Heat*, p. 106.
21. Masdar Media Centre, 'Masdar Announces Major Progress on Future Energy Strategy' (22 Jan 2008). See note 104, where the BBC quotes the cost at $22 billion, which appears to be an update on the $15 billion quoted by Masdar.
22. 'Work starts on Gulf "green city"', BBC online (10 Feb 2008).

Chapter 4 Spirit of the Blitz

1. T.M.F. Smith, 'Public Opinion Polls: The UK General Election, 1992', *Journal of the Royal Statistical Society. Series A (Statistics in Society)*, 159 (1996), pp. 535–545.
2. Ipsos Mori, *Tipping Point or Turning Point? Social Marketing and Climate Change* (Aug 2007), pp. 9, 42. Also see Chris Rose et al., 'Research Into Motivating Prospectors, Settlers and Pioneers To Change Behaviours That Affect Climate Emissions' (2007).
3. www.britishenergy.com (1 Jul 2007).
4. National Statistics, 'Domestic energy consumption per household', accessed 5 Feb 2008.
5. Joint Committee on the Draft Climate Change Bill, *Draft Climate Change Bill*, (HMSO, London, 2007).
6. Stop Climate Change Scotland, 'SCCS Response to Climate Bill Consultation Launch' (5 Feb 2008).
7. Scottish Government, 'Scotland's Climate Change Programme: Annual Report 2007' (Mar 2007).

8. Roger Bate and Julian Morris, 'Global Warming: Apocalypse or Hot Air?' (Institute of Economic Affairs, 1994).
9. 'Blair green views "muddle headed"', BBC online (9 Jan 2007).
10. 'Have Your Say: 'Can aircraft be environmentally friendly?' BBC online (9 July 2007).
11. 'Have Your Say: 'Would you be willing to pay green taxes on short-haul flights?' BBC online (13 September 2007).
12. Artyom Liss, 'Global warming leaves Russians cold', BBC online (4 Sep 2007).
13. David Ross, 'Why Scotland can be a top wine-maker . . . in 80 years', *The Herald* (27 Aug 2007).
14. Brian Donnelly, 'Nairn: thank climate change for Scots food', *The Herald* (12 Feb 2008), my emphasis.
15. Iain MacWhirter, 'Scotland could play lead role on global warming', *Sunday Herald* (4 Feb 2007).
16. Richard Sadler, 'Roads to Ruin', *The Guardian* (13 Dec 2006).
17. David Millward, 'Bus and train fares zoom past car costs', *The Telegraph* (4 Dec 2007).
18. David Prest, 'Evacuees in World War Two – the True Story', BBC online (5 Feb 2008).
19. Jon Kelly, 'You've got to show the blitz spirit', BBC online (22 Jul 2007).
20. Evan Davis, 'Cost of the floods', BBC online (25 Jul 2007).
21. *The Week*, 625 (4 Aug 2007), p. 40.
22. Charles Tart, *Waking Up: Overcoming the Obstacles to Human Potential* (Shambhala, Boston, 1986).
23. Tuvalu TIDC, 'Tuvalu and global warming' (8 Dec 2007).

PART 2

Chapter 5 – Pride and Ecocide

1. Georgius Agricola, *De Re Metallica* (Basel, 1556), Book 1, trans. H.C. Hoover and L.H. Hoover (London, The Mining Magazine, 1912), pp. 6–7.
2. John G. Neihardt, *Black Elk Speaks: Being the Life Story of a Holy Man of the Oglala Sioux*. (Bison Books, Lincoln, Nebraska, 2000), p. 4, my emphasis.
3. Genesis 10:8–12; 11:1–9.
4. Flavius Josephus, *Antiquities of the Jews,* Part I, chapter iv, 2, (FullBooks.com, 7 Oct 2007).
5. See Walter Wink, *The Powers that Be: Theology for a New Millennium* (Doubleday, New York, 1998).
6. Psalms 127:1.
7. Smithsonian Institution Human Origins Program, 'Homo Sapiens' (March 2006).
8. Clive Ponting, *A Green History of the World: The Environment and the Collapse of Great Civilizations* (Penguin, London, 1993), pp. 69–73 citing C. Leonard Woolley's *Ur of the Chaldees*.
9. Ibid.
10. See Marija Gimbutas, *The Goddesses and Gods of Old Europe* (Thames and Hudson, London, 1982).
11. See Ronald Hutton, *The Pagan Religions of the Ancient British Isles* (Blackwell, Oxford, 1993).
12. Martin Teicher, 'Scars that won't heal: the neurobiology of child abuse', *Scientific American* online edition (Mar 2002), p. 20.
13. Kate Cairns, *Attachment, trauma and resilience: therapeutic caring for children* (British Association for Adoption and Fostering, London, 2002); James Gilligan, *Violence: Reflections on a National Epidemic* (Vintage, New York, 1997); Alice Miller, *For Your Own Good: The Roots of Violence in Child-rearing* (Virago, London, 1987).
14. N.K. Sandars, trans., *The Epic of Gilgamesh* (Penguin Classics, Harmondsworth, 1960).

15. For Biblical geopolitics of cedars of Lebanon, see Michael Northcott, *A Moral Climate: the ethics of global warming* (Darton, Longman and Todd, London, 2007), pp. 101–7.
16. Job 1:7 (NRSV).
17. Plato, *Timaeus and Critias,* trans. Desmond Lee (Penguin Classics, Harmondsworth, 1965), 110–12.
18. Plato, *Critias*, 120–1.
19. Plato, *Timaeus*, 25.
20. A.E. Taylor, trans., 'Laws' in Edith Hamilton and Huntington Cairns, eds., *The Collected Dialogues of Plato* (Princeton University Press, Princeton, NJ, 1961), Book 3, 676–81.
21. J.B. Skemp, trans., 'Statesman' in Edith Hamilton and Huntington Cairns, eds., *The Collected Dialogues of Plato* (Princeton University Press, Princeton, NJ, 1961), 268–76.
22. Gordon Campbell, 'Empedocles (of Acragas)', Internet Encyclopedia of Philosophy (2006). Also Robin Waterfield, *The First Philosophers: The Presocratics and the Sophists* (Oxford University Press, Oxford, 2000), pp. 133–63.
23. Waterfield, *The First Philosophers*, p. 150.
24. Campbell, 'Empedocles', Internet Encyclopedia of Philosophy (2006).
25. Lovelock, *Revenge of Gaia*, pp. 53–54.
26. Kevin Edwards and Ian Ralston, 'Environment and People in Prehistoric and Early Historical Times: Preliminary Considerations', pp. 1–10; and Colin K. Ballantyne and Alastair G. Dawson, 'Geomorphology and Landscape Change', pp. 23–44, both in Kevin Edwards and Ian Ralston, eds., *Scotland After the Ice Age* (Polygon, Edinburgh, 2003).
27. Vivien Gornitz, 'Sea Level Rise, After the Ice Melted and Today', Science Briefing, NASA – Goddard Institute for Space Studies (Jan 2007).
28. James Hansen, 'Huge sea level rises are coming – unless we act now', *New Scientist* 2614 (25 Jul 2007), pp. 30–34.

29. Werner Nutzel, 'On the Geographical Position of as Yet Unexplored Early Mesopotamian Cultures: Contribution to the Theoretical Archaeology', *Journal of the American Oriental Society* 99:2 (1979), pp. 288–96.
30. K. Lambeck, P. Johnston et al, 'Late Pleistocene and Holocene sea-level change', *Geodynamics – Extract from RSES Annual Report*, Australian National University (1995). Also 'Persian Gulf Once Dry, Green, and Inhabited by Humans: Implications', SEMP *Biot Report* 422 (15 May 2007).
31. Ballantyne and Dawson, 'Geomorphology and Landscape Change' in Edwards and Ralston, eds, *Scotland After the Ice Age*, p. 39.
32. Ibid., p. 35.
33. Harsh K. Gupta, 'Artificial water reservoir-triggered earthquakes with special emphasis at Koyna', *Current Science* 88:10, pp. 1628–31 (25 May 2005).
34. Michael Le Page's footnote to Hansen, 'Huge sea level rises are coming – unless we act now', *New Scientist* (25 Jul 2007).
35. William F. Ruddiman, 'How did Humans first alter Global Climate?', *Scientific American*, 292:3 (March 2005), pp. 46–53.
36. Isaiah 24:4–6 (NASB; one of the most literal translations).

Chapter 6 – Dissociation of Sensibility

1. The World Commission on Environment and Development, *Our Common Future* (Oxford University Press, Oxford, 1987), p. 19.
2. Plato, *The Republic,* trans. A.D. Lindsay (Everyman, London, 1936), II:372–74.
3. William Shakespeare, *Macbeth* (Penguin, London, 1995), I, vii, 27 (vaulting); II, i, 49–50 (half-world); II, iv, 8 (make war); II, iv, 5 (troubled).
4. *Macbeth*, IV, i, 42–43.

5. Lawrence Normand and Gareth Roberts, *Witchcraft in Early Modern Scotland: James VI's* Demonology *and the North Berwick Witches* (University of Exeter Press, Exeter, 2000), p. 80.
6. Thanks to historian Jim Hunter for pointing this out on the boat to Eigg, 12 June 2007.
7. Normand and Roberts, *Witchcraft in Early Modern Scotland*, p. 246.
8. 'Newes from Scotland', full annotated text in Normand and Roberts, *Witchcraft in Early Modern Scotland*, p. 318.
9. Edward H. Thompson, 'Bothwell and the North Berwick Witches: A Chronology', accessed 26 Feb 2008.
10. Edward Cowan, 'The Darker Vision of the Scottish Renaissance: The Devil and Francis Stewart' in Brian Levack, ed., *Witchcraft in Scotland (Articles on Witchcraft, Magic and Demonology, Vol 7)*, (Routledge, London, 1992), p. 234.
11. 'The Survey of Scottish Witchcraft', Department of Scottish History, Edinburgh University (2005).
12. Pers. com., 11 Aug 2007.
13. Cited in draft of Michael Newton's book on Scottish Highland ontology, forthcoming from Birlinn (2009).
14. Verbally from St Bride's Day lecture by John MacInnes, Edinburgh University, 1 Feb 1996. See also his magisterial papers: Michael Newton, ed., *Duthchas Nan Gaidheal: Collected Essays of John MacInnes* (Birlinn, Edinburgh, 2006).
15. Nan Shepherd, 'The Living Mountain' in *The Grampian Quartet* (Canongate Books, Edinburgh, 1996), Book 4, p. 84.
16. Alastair McIntosh, 'Faerie Faith in Scotland' in J. Kaplan and B. Taylor, eds., *Encyclopedia of Religion and Nature* (2 vols., Continuum International Publishing, 2005).
17. W.B. Yeats, *The Celtic Twilight* (reprint, Prism Press, Bridport, 1990, originally 1893), pp. 92–95.
18. Martin Scofield, *T.S. Eliot: The Poems* (Cambridge University Press, Cambridge, 1988), p. 10.

19. T.S. Eliot, 'The Metaphysical Poets', *Times Literary Supplement* (20 Oct 1921).
20. Iain Crichton Smith, 'Poetic Energy, Language and Nationhood' in *Towards the Human: Selected Essays* (Macdonald Publishers, Loanhead, 1986), p. 87. (Smith says he echoes Edwin Muir in this opinion.)
21. See Iain Crichton Smith, 'The Feeling Intelligence' and 'The Golden Lyric' in *Towards the Human.*
22. C.G. Jung, *Collected Works Vol. 17: The Development of Personality* (Routledge, London, 1992).
23. Thomas Gray, 'Elegy Written in a Country Churchyard' (1751).
24. T.S. Eliot, 'The Hollow Men' (1925).
25. G. Gregory Smith, *Scottish Literature* (Macmillan, London, 1919), pp. 20, 19, 40.
26. Hugh MacDiarmid, 'The Caledonian Antisyzygy and the Gaelic Idea' (1931–32) in Duncan Glen, ed., *Selected Essays of Hugh MacDiarmid* (Jonathan Cape, London, 1969), pp. 56–74.
27. Rudyard Kipling, *Puck of Pook's Hill* and *Rewards and Fairies* (Oxford University Press, Oxford, 1993), p. 14.
28. See W.Y. Evans-Wentz, *The Fairy Faith in Celtic Countries* (Citadel Press – Carol Publishing, New York, 1994, first published in 1911). Lizanne Henderson and Edward J. Cowan, *Scottish Fairy Belief: A History* (Tuckwell, Edinburgh, 2001); John Gregorson Campbell, *The Gaelic Otherworld: Superstitions of the Highlands and Islands of Scotland and Witchcraft and Second Sight in the Highlands and Islands* with an introduction by Ronald Black (Birlinn, Edinburgh, 2005); John MacInnes, 'Looking at Legends of the Supernatural' in Michael Newton, ed., *Selected Essays of John MacInnes* (Birlinn, Edinburgh, 2006), pp. 459–76.
29. 'On Poetry and Poets' cited in Martin Scofield, *T.S Eliot: The Poems* (Cambridge University Press, Cambridge, 1988), p. 78.

Chapter 7 – Colonised by Death

1. John Kenneth Galbraith, *The Affluent Society* (Mariner Books, New York, 1998), pp. 128–29.
2. Clive Hamilton and Richard Denniss, *Affluenza: When Too Much is Never Enough* (Allen and Unwin, Sydney, 2005), pp. 6–7.
3. TIME, 'Associated Advertising Clubs of the World' (25 May 1925). Emphasis mine.
4. Edward Bernays, *Propaganda* (Ig Publishing, New York, 2004), p. 37.
5. Ibid., pp. 71, 84, 39.
6. Ibid., pp. 75, 76.
7. See David Miller and William Dinan, *A Century of Spin: How Public Relations Became the Cutting Edge of Corporate Power* (Pluto, London, 2008).
8. Vance Packard, *The Hidden Persuaders* (Penguin, London, 1960), pp. 27–28.
9. Carl Jung, *Memories, Dreams, Reflections* (Collins, Glasgow, 1967), p. 56.
10. Carl Jung, *Man and his Symbols* (Picador, London, 1978), p. 84.
11. See Dichter Ernest, 'Why Do We Smoke Cigarettes?', *The Psychology of Everyday Living* (Barnes & Noble Inc., New York, 1947).
12. Packard, *Hidden Persuaders,* pp. 193–5.
13. André Breton, *Manifestoes of Surrealism* (University of Michigan Press, Ann Arbor, 1972), pp. 26, 14. This translation refers to 'thought' rather than 'mind', but I have substituted the word most widely quoted.
14. Sigmund Freud, 'The Ego and the Id', *On Metapsychology* (Penguin, London, 1984), pp. 380–81.
15. On Bernays, see Michael E. Jones, 'The Torches of Freedom Campaign: Behaviorism, Advertising, and the Rise of the American Empire', (1999), CultureWars.com; on Dichter, see Ernest Dichter, 'Why Do We Smoke Cigarettes?', *The Psychology of Everyday Living* (Barnes & Noble Inc., New York, 1947).

16. Richard Rohr, *Adam's Return: The five promises of male initiation* (Crossroad, New York, 2004), pp. 140–42.
17. The Beatles, 'Let it Be' from the album of the same name (Apple, 1970), quoting Luke 1:38 as in the English Standard Version.
18. Joseph Campbell, *The Hero With A Thousand Faces* (Fontana, London, 1993), p. 40.
19. Vanessa Thorpe, 'Culture Vulture', *The Observer* (2 Dec 2001).
20. Stuart Jeffries, 'What Charles did next', *The Guardian* (6 Sep 2006).
21. 'Transcript of President Dwight D. Eisenhower's Farewell Address', OurDocuments.gov (17 Jan 1961). This is also available here: Andrew Burnet, ed., *Chambers Book of Speeches* (Chambers Harrap, Edinburgh, 2006), pp. 286–89.

Chapter 8 – Journey into the Soul

1. Ezekiel 28 (NRSV).
2. Land Rover, 'Introduction to the Range Rover', accessed 24 Jan 2008.
3. Arthur Deikman, 'Deautomatization and the Mystic Experience', *Psychiatry* 29 (1966).
4. Audre Lorde, 'The Uses of the Erotic: The Erotic as Power' in *Sister Outsider: Essays and Speeches* (Crossing Press, Berkeley, CA), 1984, pp. 53–59.
5. See Eliot, *The Metaphysical Poets*.
6. Walter Wink, *Engaging the Powers: Discernment and Resistance in a World of Domination* (Fortress Press, Minneapolis, MN, 1992).
7. Plato, *The Last Days of Socrates – Euthyphro, The Apology, Crito, Phaedo,* trans. Hugh Tredennick (Penguin, Harmondsworth, 1959).
8. IUCN, 'Red List of Threatened Species', accessed 28 Feb 2008.
9. IPCC, *AR4 SYR*, pp. 13, 21.
10. US Census Bureau, 'Historical estimates of world population' (16 Jul 2007).

11. R.B. Strothers, 'Mystery cloud of AD 536', *Nature* 304, pp. 344–345 (26 Jan 1984).
12. Wikipedia, 'Extreme weather events of 535–536', accessed 26 Jan 2008.
13. Arno Karlen, *Plague's Progress: A social history of man and disease* (Phoenix, London, 2001), p. 145.
14. World Health Organisation 'Cumulative Number of Confirmed Human Cases of Avian Influenza A/(H5N1) Reported to WHO' (24 Jan 2008).
15. Robert G. Webster and Elizabeth Jane Walker, 'Influenza – The world is teetering on the edge of a pandemic that could kill a large fraction of the human population', *American Scientist* 91 (Mar–Apr 2003).
16. Asian Development Bank, Asian Development Outlook, 'Assessing the Impact and Cost of SARS in Developing Asia' (2003).
17. World Health Organisation, 'WHO checklist for influenza pandemic preparedness planning', (2005).
18. T.S. Eliot, 'Four Quartets – (East Coker)', 1935–42.
19. See Elisabeth Kübler-Ross, *On Death and Dying* (Simon and Schuster, London, 1997).
20. Joanna Macy, 'The Great Turning' (2007).
21. Oswald Chambers, 'The Discipline of Disillusionment', SingleVision Ministries (2008).
22. Tyndall Centre, 'Made in China: Who is responsible for China's rapidly rising CO_2 emissions . . .' (19 Oct 2007).
23. 'Robert Burns' and 'James Hutton', *Oxford Dictionary of National Biography*, online: www.oxforddnb.com. (Thanks to Professor Stuart Haszeldine of Edinburgh University's geology department for this.)
24. SCRAN, Print entitled 'Burns' first meeting with Scott,' undated (thanks to Ruth Watkins of Stirling University for this and also Sandy Hutchison, a Burns scholar at the University of the West of Scotland).
25. Keith Stewart Thomson, 'Vestiges of James Hutton', *American Scientist Online* 89 (May–Jun 2001), my emphasis quoted from the original 1785 version.

26. James Hutton, *Theory of the Earth* (1795), Book 1, Chapter 1.
27. Cited in Hutton's entry in the *Oxford Dictionary of National Biography*.
28. Brian Keenan, *Turlough* (Vintage, New York, 2001), p. 3.
29. Olivier Clement, *The Roots of Christian Mysticism* (New City Press, New York, 1993), p. 60.
30. Jeremiah 12:4 and 32 (NRSV). See also Walter Bruggemann, *Hopeful Imagination: Prophetic Voices in Exile* (Fortress Press, Minneapolis, MN, 1986).
31. 1 Corinthians 13.
32. Campbell, John Lorne and Hugh Cheape (ed.), *A Very Civil People: Hebridean Folk, History and Tradition* (Birlinn, Edinburgh, 2004), p. 109.
33. Deep Purple, 'Child in Time', *Deep Purple in Rock* (1970).

Chapter 9: Towards Cultural Psychotherapy

1. Plato, *Republic*, trans. A.D. Lindsay, VII:515–19.
2. Ibid., VII:517.
3. My warm thanks to Brian Pearce for his permission to step aside from Chatham House rules.
4. The Bishop of Liverpool, the Right Rev James Jones, pers. com. (2007).
5. Kevin MacNeil, *The Stornoway Way* (Penguin, London, 2006), pp. 232–33.
6. Carl Jung, *Memories, Dreams, Reflections* (Fontana, Glasgow, 1967), p. 190.
7. See Wink, *Engaging the Powers*.
8. Cairns, *Attachment, trauma and resilience*, p. 102.
9. Stanislav Grof and Christina Grof, eds., *Spiritual Emergency: When Personal Transformation Becomes a Crisis* (Tarcher, New York, 1989); Roger Walsh and Frances Vaughan, eds., *Paths Beyond Ego: Transpersonal Vision* (Tarcher/Putnam Books, New York, 1993).
10. See Jack Kornfield, *A Path With Heart* (Rider & Co., London, 2002).

11. Gilligan, *Violence,* p. 47.
12. Camara's full text is in PDF at www.alastairmcintosh.com/general/spiral-of-violence.htm.
13. Walter Wink, ed., *Peace is the Way: Writings on Nonviolence from the Fellowship of Reconciliation* (Orbis, Maryknoll, NY, 2000).
14. Manfred Max-Neef, 'Development and Human Needs' in Paul Ekins and Manfred Max-Neef, eds., *Real-life Economics: Understanding Wealth Creation* (Routledge, London, 1992), pp. 197–214.
15. Kevin Roberts, *Lovemarks: the future beyond brands* (Powerhouse, London, 2006), pp. 36, 57 and back cover. See also Jane Kendall and Tom Crompton, 'All you need is love to protect the natural world', *The Guardian* (27 Feb 2008).
16. See 'Weathercocks' www.valuingnature.org.
17. Fairtrade Foundation, 'Seven million farming families worldwide benefit as global Fairtrade sales increase by 40% and UK awareness of the Fairtrade Mark rises to 57%', press release (10 Aug 2007).
18. See Theodore Roszak, Mary E. Gomes and Allen D. Kanner, eds., *Ecopsychology: Restoring the Earth – Healing the Mind* (University of California Press, Berkeley, 2002).
19. See Ernest Renan, Lecture at Sorbonne, 'What is a Nation?' (*Qu'est-ce qu'une nation?*), 11 Mar 1882. Available online, but also in Geoff Eley and Ronald Grigor Suny, eds., *Becoming National: A Reader* (Oxford University Press, New York and Oxford, 1996), pp. 41–55.
20. Alastair McIntosh, 'Scotland' in *Love and Revolution (collected poetry)* (Luath Press, Edinburgh, 2006), p. 56.
21. For example, 500 groups across Scotland that participated in the 'People and Parliament' process, 1999. This can be found online.
22. See www.GalGael.org.
23. Iain Crichton Smith, 'Real People in a Real Place' in *idem.*, *Towards the Human: Selected Essays* (Macdonald Publishers, Loanhead, 1986), pp. 13–70. Available online.

24. David Orr, 'The Rhythm of Gratitude', *Resurgence* 247 (Mar/Apr 2008), pp. 10-11.
25. Nikolai Berdyaev, *The Paradox of the Lie* (1939), trans. Fr. S. Janos (2000), The Berdyaev Online Bibliotek Library, www.berdyaev.com, accessed 3 Mar 2008).
26. Leo Tolstoy, *Anna Karenina*, trans. Pevear and Volokhonsky (Penguin, London, 2000), Part 8.

Afterword

1. European Space Agency, 'Satellites witness lowest Arctic ice coverage in history' (14 Sep 2007).
2. Matthew 5:13.

INDEX